'A Veritable Eden'

Entrance lodge of the Manchester Botanic Garden.

'A Veritable Eden'

The Manchester Botanic Garden: a History

by Ann Brooks

WIND*gather*
PRESS

This book is dedicated to
Pat Hartwell and Hélène Williams

They taught me much about gardening and more about enjoying life

Stately spires of Delphiniums and Campanulas, painted by Josephine Gundry.
W. P. WRIGHT, *HARDY PERENNIALS AND HERBACEOUS BORDERS* (LONDON, 1912), 84.

Windgather Press
is an imprint of
Oxbow Books, Oxford

ISBN 978-1-905119-37-0

A CIP record for this book is available from the British Library

This book is available direct from

Oxbow Books, Oxford, UK
(Phone: 01865-241249; Fax: 01865-794449)

and

The David Brown Book Company
PO Box 511, Oakville, CT 06779, USA
(Phone: 860-945-9329; Fax: 860-945-9468)

or from our website

www.oxbowbooks.com

Printed in Great Britain by
Information Press, Eynsham, Oxfordshire

Contents

List of Illustrations

All photgraphs by the author unless otherwise stated.
Colour drawings coloured by Lizzie Holiday.

Abbreviations

....

Chronicle = *Manchester Chronicle*
City News = *Manchester City news*
Guardian = *Manchester Guardian*
MBH = Manchester Botanical and Horticultural Society

Acknowledgements

Gardening had been my passion for many years: to be able to research this lost garden was a privilege and a pleasure. Professor Hannah Barker, Department of History, and Alan Ruff, Department of Landscape and Planning (retired), both University of Manchester, enthusiastically supervised my doctorial research and supported me through this publication process, thank you both. Hannah convinced me that you are never too old to start a new adventure. My family Jane Anson, Michael Brooks, Catherine Hughes, and John Tomczyk gave unfailing support. Thanks are due to friends Vivienne Blackburn, Margaret Gill, Caroline Holmes, Phillipa Rakusen and Evelyn Taylor as well as all others who gave their help in furtherance of my quest. A special thank you is due to my former co-author, Bryan Haworth, who gave me his unstinting analysis and insightful criticism on this my first solo work. Dr Pat Sherwood brilliantly proofread the original manuscript. Terry Wyke generously gave help on the mapping of Manchester. I would like to extend my thanks and appreciation for the support and encouragement I received from John Milner and the Committee of the Royal Botanical and Horticultural Society of Manchester and the Northern Counties.

I am very grateful to the staff of all the Archives and Libraries credited throughout the text. Special thanks go to Mike Powell and Fergus Wilde, Chetham's Library, Manchester; Stella Butler and the staff at Special Collections, John Rylands University Library, Manchester; Emma Marigliano, The Portico Library, Manchester; and Graham Hardy and the staff of The Library, The Royal Botanic Garden, Edinburgh. Thanks also go to the many archivists and Curators who looked to see if their collection contained Findlay's *Guide to the Manchester Botanic Garden 1857* [still missing].

I offer my particular appreciation to my editor, Dr Julie Gardiner, and to the staff at Oxbow Books who made the publishing process virtually painless, thank you all. Finally, but certainly not least, I thank my husband Egan, who served as chauffer, confidante and general factotum, for his support and encouragement when his wife began another career.

Preface: 'A Veritable Eden'[1]

The Manchester Botanic Garden opened in 1831 and, after a chequered history including national fame and financial disaster, it in essence ceased to exist in 1908. The Society that founded it survived and still continues to work to the benefit of botany and horticulture in Manchester and the northern counties. Few in Manchester now know that the city was once home to a Botanic Garden with an international reputation: a garden that hosted some of the most prestigious horticultural exhibitions in the latter half of Victoria's reign. The one remaining legacy is the façade is illustrated in the frontispiece. It stands forlorn on White City Way within a mile of the city centre and now graces the frontage of the White City Shopping Centre (2009). Some Mancunians remember when it was the entrance gate to a greyhound-racing stadium or have heard of the White City amusement park but the Garden itself is forgotten. This book is a history of that Garden and the Society that founded it.

Why should a vanished garden be a subject of significance, not only to Manchester but to the country as a whole? The Manchester Society and its garden was a prime example of a specific type of nineteenth-century endeavour: the subscription botanic garden. This was a national movement that aimed to bring the study of botany and horticulture within the reach of the subscribing members by adding a prestigious botanic garden to their town.[2] Gardens were founded in towns across the country, including Manchester. Yet, despite their prominence at the time, we know very little about them today. The gardens, which were established to enable the scientific study of botany and horticulture,

Former Entrance Gates to the Manchester Botanic Garden

Entrance to the Manchester Botanic Garden 1831. From *The Mirror of Literature, Amusement and Instruction* 536, vol. 19, Saturday 3rd March 1832, 129.
AUTHOR'S COLLECTION.

were started by local societies with a membership that paid subscriptions to join. The use of subscriptions to finance prestigious public ventures was an accepted method by the nineteenth century; in this way an array of influential artistic and scientific institutions brought fame to the town and honour to the enlightened citizens who sponsored them. The scale and form of each garden was influenced by both fashion and the dictates of the subscribers. The provincial subscription garden movement began in Liverpool in 1803, and by 1836 several such societies and their gardens had been founded in cities around Britain.[3] The development of subscription botanic gardens in Liverpool (1803), Hull (1812), Glasgow (1817), Birmingham (1832), and Sheffield (1836) are those used throughout the book as counterpoints in this history of the Manchester Garden.[4]

Botanic gardens have a long history and it is important to see the Manchester Garden in the context of this tradition as these subscription gardens differed in several respects from those that had preceded them. From the medieval monastery herbal garden the botanic garden progressed to the Renaissance Universities of Padua and Pisa where it became a medical teaching garden. In England the first such university botanic garden was at Oxford, founded in 1621. The Society of Apothecaries founded their Physic Garden at Chelsea in 1673, initially to teach their students to recognise medicinal plants. Moving away from the universities, the Royal Botanic Garden, Kew, founded in 1753 by Princess Augusta, mother of George III, followed in the tradition of private botanic gardens established by the aristocracy throughout Europe.

The nineteenth-century subscription gardens followed the tradition of having botanical beds and horticultural display areas. In addition membership gave them access to exotic plants grown in the garden and hothouses, many of which came from overseas contacts. They could therefore ornament their own gardens with the latest introductions to the country. The subscription gardens fulfilled yet another purpose for their members and Manchester was no exception. By the 1830s they had all become pleasure gardens, as well as botanical gardens, used

The site of the rear gate to the Manchester Botanic Garden at the end of Botanical Avenue, Old Trafford, Manchester.

for social intercourse and as an escape from the growing industrial problems posed by the crowded industrial towns and the resulting pollution. This role as a pleasure garden was to become more and more important to the membership and have a significant effect on the structure and use of the gardens as the century progressed.

In designing the subscription gardens as partly a pleasure garden the founders could draw on the tradition of the English landscape movement. This influence is evident in the integration of arboreta, lakes, rockeries, and glasshouses, the latter having the dual purpose of botanical and fashionable display. Manchester, in common with the other Societies, had aristocratic patrons and, certainly on their foundation, such patrons were sources of both advice and plant material for the members. They also added a social cachet that was an additional attraction for the membership. However it will be seen that there was a significant difference from the aristocratic gardening milieu. Within the subscription gardens, there was no reference to the antique: no temples or grottos, no statues of classical gods though, for other reasons, some entrance gates were an exception to this rule. The Manchester garden reflected the subscribing members' cultural assumptions that, as the new leaders of the town, they looked to the future not the past. The Manchester garden reflected the members' politics, their interest in botany and the applied sciences, firmly placing the garden as a place to express their religious beliefs.

Notes

1   A quotation from *The Mirror of Literature, Amusement and Instruction* 536, Vol. 19, Saturday 3 March 1832, 129 (Archives of the MB H, MBH 7/3/1); John Rylands Library, Manchester.

2   In 1778 William Curtis, who had founded *The Botanical Magazine* in 1787, circulated a proposal for the opening by subscription of his London Botanic Garden. This garden was probably the first subscription garden opened in Britain and subscribing members could study the plants and exchange ideas.

3   A. Brooks, *A Veritable Eden: the Manchester Botanic Garden 1827–1907 and the movement for subscription botanic gardens.* unpublished PhD thesis, Manchester University, 2007.

4   Others were founded in non-industrial towns, such as Bury St Edmunds (1819), York (1822), and Bath (1845); only those in industrial cities are discussed in the text.

CHAPTER ONE

Manchester 1822

Market Street, Manchester 1822. Frontispiece from A. Darbyshire, *A Booke of Old Manchester and Salford* (Manchester, 1887).
AUTHOR'S COLLECTION

On 5 July 1822 a letter from Miss Harriet Hyacinth appeared in the *Manchester Iris*.[1]

> Pray, Sir, what is this Botanical Garden that is so much the subject of conversation at present?

It was probable that she had been following the series of letters published weekly in June in the same newspaper.

A lively debate had begun, on 8 June, with a letter from a pseudonymous author 'A Botanist'.[2] He lamented that Manchester did not possess a botanic garden like 'the most beautiful Botanic Garden of our neighbours Liverpool'. He knew of several who were in favour of such a resort. It would appeal to horticulturists like himself as well as to the general subscriber. In addition, it would be a most suitable promenade for the ladies of the town as they only had St Ann's Square and the Infirmary gardens, both of which were noisy public spaces. He added that Leeds was already planning such a garden.

An Aside: The Liverpool Subscription Botanic Garden

The first provincial subscription botanic garden was proposed in Liverpool in 1790 by a passionate botanist and polymath, William Roscoe. The Liverpool garden opened in 1803 on a 5 acre (*c.* 2 ha) site near the city centre. Liverpool's influence on the movement for subscription botanic gardens is demonstrable: Hull, founded in 1811, was laid out by John Shepherd, Liverpool's curator and in 1817 Glasgow's proposal closely mirrored Liverpool's. Shepherd was consulted by the founders of Manchester and Birmingham Botanic Gardens; all opened in the 1830s on much larger sites reflecting the subscribers desire for a pleasure ground as well as a scientific garden. By the 1830s Liverpool's garden was facing problems; the prevailing saline winds, the expansion of the town and the resulting industrial pollution. The subscribers agreed to resolve their dilemma by moving the site of the Garden to an 11 acre (*c.* 4.4 ha) plot on the south side of what was still rural Edge Hill.[1] Shepherd's design incorporated walks, lawns and a larger, spacious conservatory, an acknowledgement that a pleasure garden was now a necessity.

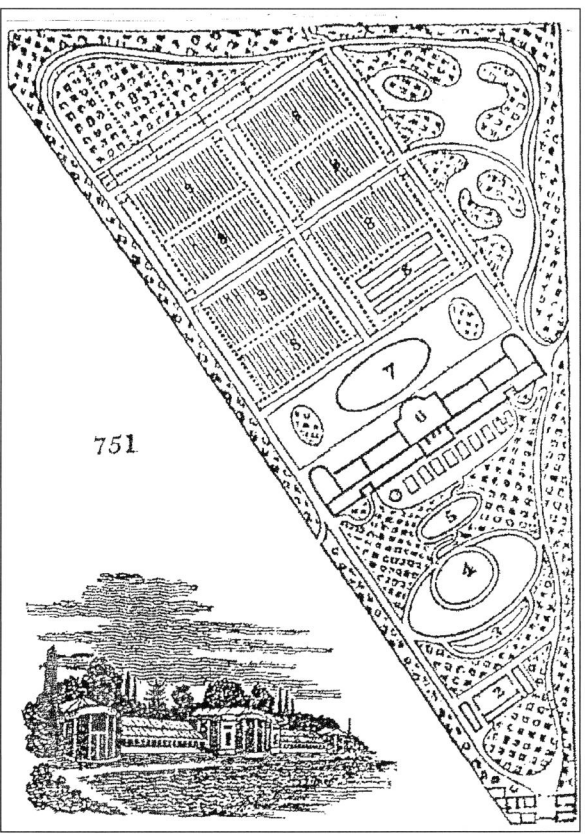

The Liverpool Botanic Garden. From J. C. Loudon, *An Encyclopaedia of Gardening New Edition* (London, 1834). Key: 1. Entrance Lodges; 2. Stove; 3. Rock plants; 4. Bog plants; 5. Greenhouse ground; 6. Conservatory; 7. Pond; 8. Herbaceous; 9. Grasses.
AUTHOR'S COLLECTION

1   The Register of Parks and Gardens of Special Historic Interest, Wavertree Botanic Garden, GD 2593, gives 1831 as the date of the decision to move and 1836 as the date of the opening of the new garden

On 15 June came a reply from an opposing citizen, 'Cop-Twist', (though the editor pointed out that the *Iris* fully supported the proposal). Though written in a comic style, he raised some valid arguments. With trade distressed would the money not be better used for public charities? He objected to Liverpool being held up as an example to follow; by this time the rivalry between the two towns was well established. In his opinion Manchester men were hard working whilst Liverpool merchants were speculators and could indulge themselves in such

The Infirmary,
Dispensary, and Lunatic
Asylum, Manchester.
From S. Austin, J.
Harwood, G. And
C. Pine, *Lancashire
Illustrated* (London,
1832).

fancies as a botanic garden. Though himself displaying an expert knowledge of plant classification, 'Cop-twist' was against the indecent study of botany by ladies as the new Linnaean system was based on the sexual characteristics of plants.[3] And he pointed out that he and his family were happy to frequent Tinker's Garden in Collyhurst.

Some of the potential subscribers may not have agreed especially as Tinker's advertising suggests that the respectable citizens would have avoided such a notorious establishment. They would have welcomed the new private garden.

On 29 June the 'Botanist's' reply was to the point. The idea for a Botanic Garden had been discussed locally for several years and was seen as a means to encourage the promotion of science, not only botany. In addition he believed, unlike 'Cop-twist', that botany was the ideal science for ladies to study as Linnaeus had classified plants using the concept of sex *within marriage* thus rendering it respectable. Botany was also utilitarian with the use of medicinal and culinary plants being of domestic importance to the ladies. 'Cop-Twist' would not back down and a week later (28 June) saw the final letter. The state of local trade was his real objection – there was no money for such a proposal. It would not reflect well on the town to do it shabbily or if the garden, being undercapitalised, failed to open. In fact Harriet Hyacinth's letter was probably a fiction, which allowed a discussion to be aired by a group known as 'The Club' who met weekly to discuss matters of importance to the town.[4] Their deliberations were then published in the *Manchester Iris*.

The members, using arguments found in the earlier letters, gave no real names but 'The President' enjoyed visiting the Liverpool garden and wanted a garden in Manchester for its agreeable walks, which allowed one to 'satisfy rational curiosity in the study of plants'. As it was to be a subscription garden

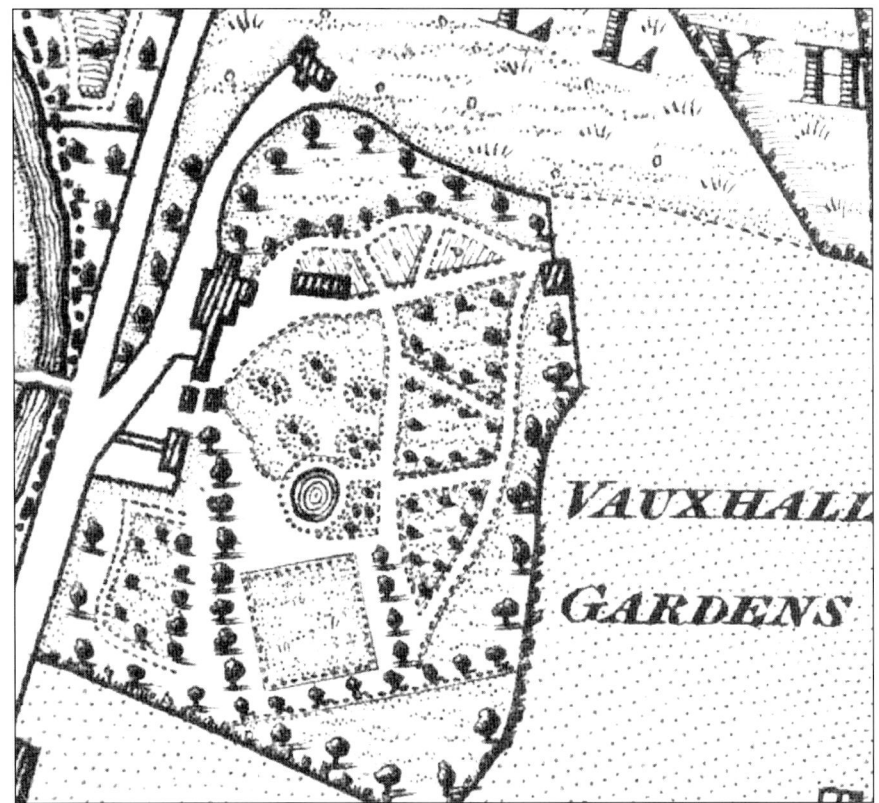

Vauxhall Gardens, Collyhurst, 1824, showing the ornamental beds and walks surrounded by trees and hedges. There are no botanical beds in evidence as this was a pleasure ground. They were also known as Tinker's Gardens after their founder Robert Tinker. Shown on Swire's Map of Manchester 1824.
CHETHAM'S LIBRARY, MANCHESTER

Vauxhall Gardens token.
AUTHOR'S COLLECTION

The commercial Vauxhall Gardens offered walks and entertainment in, Tinker claimed, a rural retreat. The Gardens were lit up in the evening and spectacles were staged. Fireworks, balloon ascents and recreations of battles were often advertised, while bands played in the gardens. Vandalism was a problem and men with dogs and firearms were employed to deter offenders who broke in to avoid payment. See Manchester Mercury, 10 September 1814.
NEWSPAPER COLLECTION, CHETHAMS' LIBRARY, MANCHESTER

it would be free of the objectionable society found in so called 'tea gardens'. 'The Doctor' extolled the pleasures of botanical studies where plants were methodically arranged. 'The Tradesman' objected to the sexual nature of this study and the 'immoral tendency of its language'. He was over-ruled as botany was seen as a cheap and innocent study combining bodily exercise with 'ardour of the mind'. Flowers charmed the sight and gave 'natural and pure delight'. It would seem however that 'Cop-Twist' spoke for the silent majority as the botanic garden was not founded. Though the proposed garden failed, the end of June 1822 saw the establishment of the Manchester Society for the Promotion of Natural History and a Manchester Floral and Horticultural Society followed two years later. The idea for a subscription botanic garden for the town was still alive however and in 1827 the establishment of a Manchester Botanic Garden was proposed again.

On 30 July 1827 a public meeting was held at Manchester Town Hall. The purpose of which was to establish 'A Manchester Botanical and Horticultural Society for the town and to construct a Botanical and Horticultural Garden for Manchester and the Neighbourhood'. Edward Loyd, the prominent Manchester banker, chaired the meeting and resolutions were passed to raise money for the project by loans and subscriptions. The local newspapers duly published the Society's aims as *An Address to the Inhabitants of Manchester and the Neighbourhood on the formation of a Botanical and Horticultural Garden*.[5] The *Address* pointed to the importance of Botany to the local population as a 'peculiarly delightful and attractive' pursuit and that Manchester was one of the foremost areas of the country where botany and horticulture enjoyed assiduous and successful study by all walks of life. Lack of money however meant that many were precluded from studying the subject especially plants from foreign countries or of great rarity. A subscription to the Society would enable all to have access to the Botanic Garden where, with the use of modern technology

Manchester Town Hall. From S. Austin, J. Harwood, G. And C. Pine, *Lancashire Illustrated* (London, 1832).

in the hothouses and conservatory, exotic plants would thrive; and subscribers could acquire them.

The final paragraph of the *Address* appealed to quite a different aspect of the garden's possible use; 'an inviting scene of public recreation' with the character of a 'fashionable resort'. The Committee proposed to design the garden with ornamental walks so that those uninterested in botany could enjoy 'the beauty of the objects, the pleasures of the society, and the animating gaiety of the scene.' A garden in fact that combined both the botanical requirements of the 'Botanist' and the features of the resort required by 'Cop-Twist' in 1822. They also claimed they would be securing a resort near the centre of town which would be free from 'the intrusion of rude familiarity and licentious mirth', another echo of the correspondence of 1822. This precept of the moral influence of botanical studies was evident to the proposers as they remarked that 'a cottager' who devoted his leisure to his garden was secured from 'the temptations to extravagance and the natural consequences of dissipated habits.'

Correspondence again appeared in the local papers. On 4 August, the day the *Address* was published, 'A Florist' maintained that the Floral and Horticultural Society had kept the need for a botanic garden alive.[6] Claiming that in wealth and power Manchester was unsurpassed, the establishment of the Botanic Garden would mean the works of Nature could be studied amongst the works of man. To this end he appealed to the Ladies of the town to support the enterprise: 'Who would be able to resist their appeal in the cause of taste and virtue?' The Ladies may have supported the enterprise as the membership lists show that, unusually for the time, women became members of the Society in

The Royal Institution Manchester. From S. Austin, J. Harwood, G. And C. Pine, *Lancashire Illustrated* (London, 1832); (original now in the Manchester City Art Gallery). One of the foundations lauded by 'A Florist'.
AUTHOR'S COLLECTION

their own right. The Editorial in the *Guardian* that same day argued that the name, Botanic Garden, might be a deterrent by suggesting the idea of science was central to its enjoyment though the *Address* had made clear that 'the beauties of Nature in all their numerous combinations' were open to all. This was a time when many of Manchester's wealthy inhabitants still lived in the inner city in Georgian terraces without gardens. On their behalf the editor wished to see a garden that was a fashionable promenade, a place of retirement, 'an agreeable retreat from the hustle and bustle of the town. Every shareholder would have a beautiful pleasure ground, which he would not prize less because it was open also to his neighbour.' The Garden was therefore to fulfil many purposes for its members. This diversity of interests was to be one of the deciding factors in the history of the Manchester Botanic Garden.

In August 1827, no time was to be lost in buying land for the garden and the meeting appointed a Committee of Management to oversee the enterprise: the capital was to come from the subscribers. The Committee voted for two classes of subscriptions, Hereditary or Life members.[7] To encourage recruitment a list of the 72 founding subscribers was then read out – they were among the most important citizens of Manchester at the time. The Patrons of the Society were The Earl of Stamford and Warrington (Dunham Massey, Altrincham), The Earl of Wilton (Heaton House, Manchester), Lord Suffield (Middleton, nr Manchester), and T. J. Trafford (Trafford Park, Manchester). The subscribers were the great and the good of the town and included bankers, lawyers, merchants, manufacturers, surgeons, physicians and local landowners. On 11 August 1827 the leader in the *Guardian* read:

> BOTANIC GARDEN – We are extremely happy to state that the subscription for the establishment of a botanic garden near this town is proceeding in a very satisfactory manner. We believe that upwards of a hundred gentlemen have put down their names as hereditary subscribers.

With such encouragement the Committee set about their task of bringing the Manchester Botanic Garden to fruition.

Notes

1   In 1822 the population of Manchester, including the inner suburbs of Ardwick, Hulme, Chorlton-on-Medlock and Cheetham had reached 126,031. Salford, its neighbour on the other bank of the Irwell, including Broughton, had 26,552 residents. Trade in the area was bad. Memories of Peterloo in 1819 were aroused when four members of the Manchester Yeomanry Cavalry were tried at Lancaster for their assault on Thomas Redford, a journeyman hatter in St Peter's Field. After a 5-day trial they were acquitted.

2   Pseudonyms were commonly used by letter-writers to the press.

3   Linnaean classification was introduced in 1735.

4   Members were both Churchmen and Dissenters. See: *The Club. A Series of Essays, Originally Published in the Manchester Iris* (Manchester, 1825).

5   This *Address* was probably also sent to interested members of the local aristocracy as a copy exists in the archives of the Earl of Stamford and Warrington of Dunham Massey who became a Hereditary Member. He also became a Life Member of the Birmingham Subscription Botanic Garden opened in 1832.

6   On the same day in an account of an Exhibition of Fruit and Flowers, the Floral and Horticultural Society appealed for support for the proposed botanic garden.

7   Hereditary members would make an immediate payment of £25 for a share and pay an annual subscription of £1 1s (1 guinea). Life members would make a payment of £10 for a share and pay an annual subscription of £2 2s (2 guineas). Each of the classes would have different degrees of entitlement. For the Hereditary member these were given as: admission to the garden for themselves, their families and strangers: their shares could be transferred by deed or by will, subject to restriction: they were eligible to sit on the management committee. The Life members were only entitled to have admission for themselves and their families,could not pass on their share and could not sit on the management committee.

CHAPTER TWO

Gardening in Manchester 1790–1887

Irlam Hall (near Manchester), Lancashire. From S. Austin, J. Harwood G. and C. Pine, *Lancashire Illustrated* (London, 1832).
AUTHOR'S COLLECTION

In 1833, Benjamin Braidley, the Boroughreeve of Manchester gave evidence to the *Select Committee on Public Walks* that there were many villas on the outskirts of Manchester. He stated that large pleasure grounds were not common; a house valued at £2000 with 2 acres [0.9 ha] of ground, would typically have rails or hedges to the street and high walls around the garden.[1] It is clear from newspaper advertising and the categories in trade directories that, by 1827 when the Society was founded, the gardening elite of Manchester already supported a commercial industry of seedsmen, nurserymen, florists, and landscape gardeners. For example John Jones of 32, Oxford Street, Manchester, supplied bulbs, fruit trees, greenhouse and stove plants, hardy herbaceous and bog-plants, and garden sundries.[2] He was also a landscape gardener, who 'devotes most of his time to laying out gardens and pleasure Grounds' and supplied honest, experienced gardeners. Jones's shop was conveniently situated between Ardwick, a wealthy desirable suburb, and Manchester town centre a mile away. As early as 1795, the wealthy of the town were building houses on the outskirts which John Aikin described in his history of the area as 'their country residences about one or two miles from the business district'.[3] He took special notice of Ardwick Green where 'the most opulent class' resided. A Mr Tipping lived in a spacious house with pleasure grounds [lower left]'.

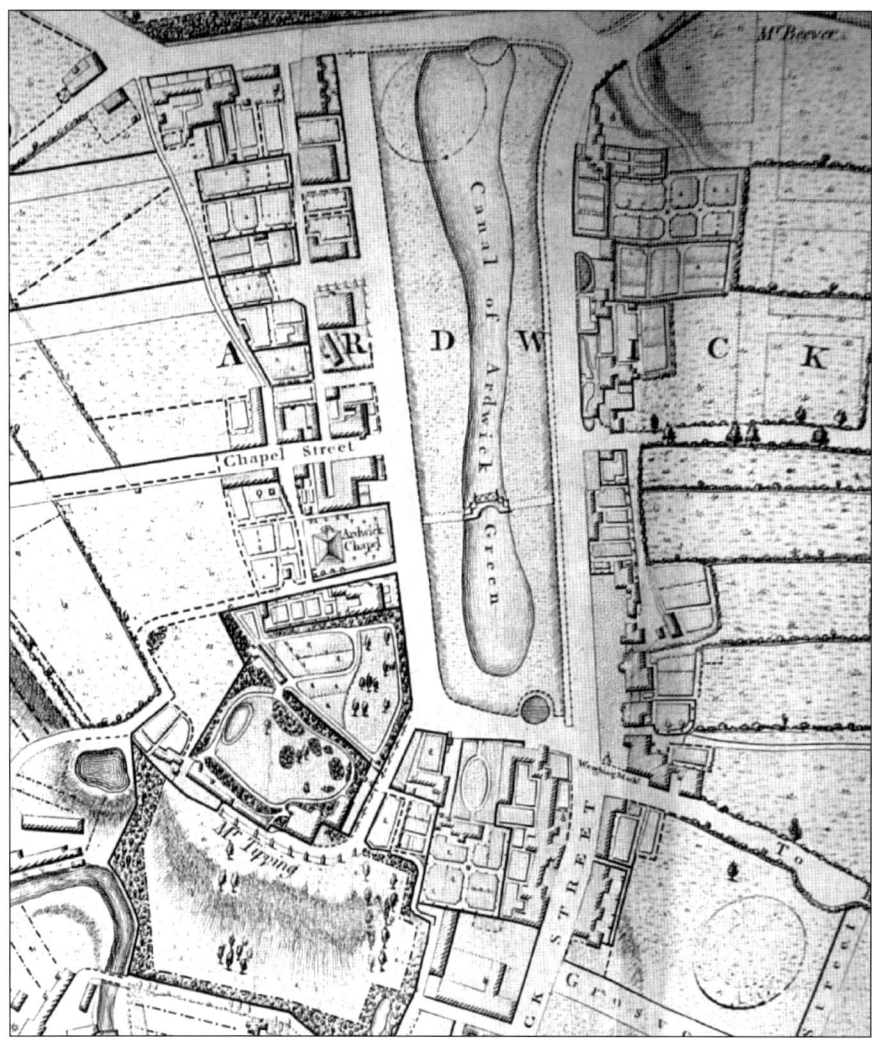

Ardwick Green shown on Laurent's *A Topographcal Plan of Manchester and Salford, by C. Laurent, Engineer* (Manchester, 1793).

Evidence from maps prepared for a late eighteenth century rating survey in the Shude Hill district of Manchester shows a variety of houses with gardens.

These plans show that many types of houses near the centre of the town had their own gardens. Larger houses incorporated more features, some of which, for example the parterre, were not purely functional but ornamental. This allowed the individual owners' gardening tastes to be displayed and can also be read as their interpretation of the fashionable gardening scene. Wealthy owners of houses in the town with no gardens could rent Summer gardens on the higher slopes of the town for pleasure and relaxation during the summer months as shown below. This practice was also common in other towns at the beginning of the nineteenth century.[4] As well as maps, newspaper advertisements are a source of information on available houses to rent or buy in the town.

Tracing the planting is more difficult. In the late eighteenth century exotic plants began to be grown not only by the aristocracy but also by the wealthy

Terraced houses and tenements with individual back gardens on Shude Hill, Manchester (Hulme Rate Documents, n.d.).
CHETHAM'S LIBRARY, MANCHESTER

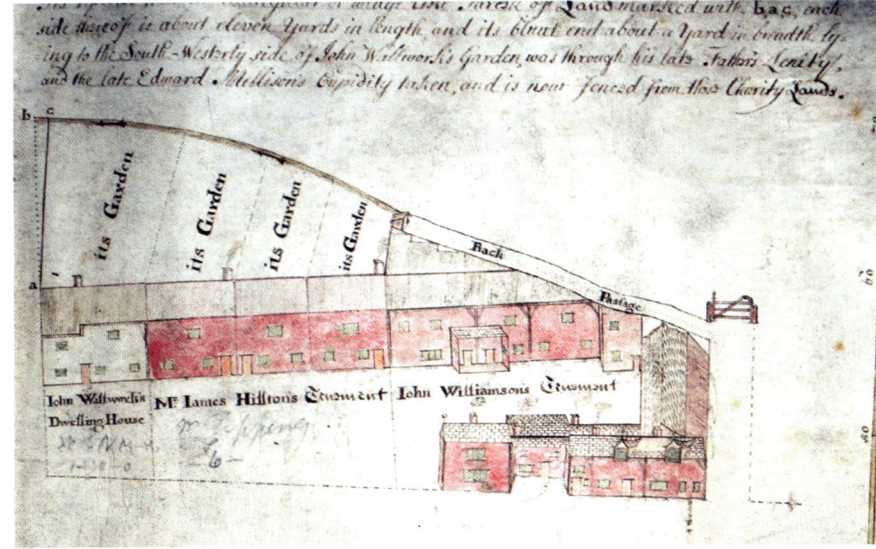

Large houses with gardens were also shown in the area; the back garden to Roger Bradshaw's tenement on the corner of Miller's Lane and Green Lane at Shude Hill Top was planted with several trees (possibly an orchard), to the front was a small parterre (Hulme Rate Documents, n.d.).
CHETHAM'S LIBRARY, MANCHESTER

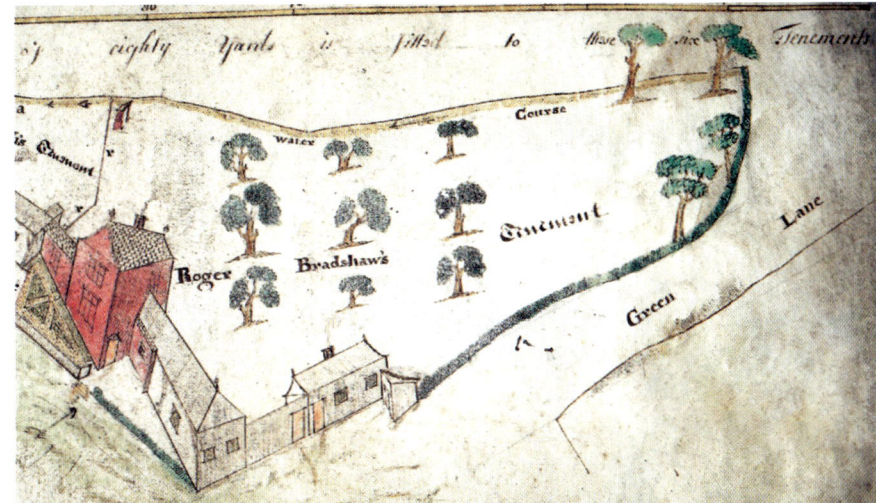

Summer gardens to rent on Shude Hill, Manchester (Hulme Rate Documents, n.d.). See J. Harding and A. Taigel, 'An air of detachment: town gardens in the eighteenth and nineteenth centuries', *Journal of the Garden History Society* (1996) 24, 238–9. St Ann's Allotments, Nottingham, founded in 1830 and still extant, were known as 'detached urban pleasure gardens'.
CHETHAM'S LIBRARY, MANCHESTER

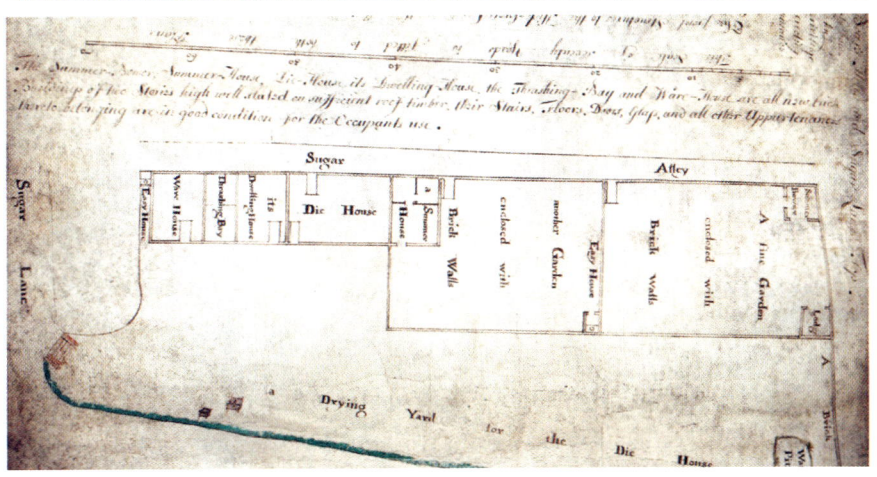

A GARDEN to be Sold or Lett,

CONSISTING of a Succeffion Houfe, a Green-houfe and Stove together, containing a great Variety of exotic Plants, likewife a good Collection of Auriculas, Carnations, and other Arbatious Plants, with a great many other Articles, viz. Shades, Glaffes, Flower Sticks and Covers, the Property of Mr. James Hallows, near St. John's Church.

For further Particulars enquire of Mr. Joseph Lockett, at the Green Man, in Jackfon's-Row.

A GARDENER Wanted, that underftands the Management of Pines, &c. Apply to Mr. James Hallows, at his Warchoufe, in Bank-ftreet, Manchefter.

Newspaper Advertisement, *Manchester Mercury*, 6 May 1798.

upper classes, who could afford to maintain hothouses. Clues can be found in the customer ledgers and daybooks of Caldwell's Nurseries, Knowsley and Knutsford, which begin in 1789. These show that, by 1827, they supplied goods not only to the aristocracy and gentry, but also to the upper classes in Manchester and environs, including to members of the Botanical Society.[5]

Wealthy Mancunians were at the forefront of this gardening revolution, as is shown by the 1779 sale catalogue of plants of a Manchester doctor, Philip Brown.[6] Brown, who had a stove and greenhouses, claimed there were plants in his collection unique to England and he thanked 'merchants in all parts of the Globe' and ships' captains for procuring 'a very large collection of foreign seeds'. Besides tender exotics the sale included herbaceous plants, bulbs, and perennials and all plants listed are priced. On 21 May 1821, the *Manchester Mercury* reported that Manchester had suffered 'a most violent thunderstorm' of heavy rain and hail. Robert J. J. Norreys of Davy Hulm Hall, magistrate and deputy-lieutenant of Lancashire, had lost over 2000 panes of glass in his hothouses, greenhouses, and pine-pits. He was an expert on pineapples but not the only one. Peter Marsland, a Manchester cotton manufacturer, contributed an article on their cultivation to *The Transactions of the Horticultural Society of London* in 1824.[7] Both were to be Botanical Society members.

The 1820s saw the upper middle class living in the city and the ring of suburbs near the centre of the town – these were where the fashionable lived. 1838 saw the publication of J. C. Loudon's *The Suburban Gardener* (1838) as, in his opinion as one of the gardening guru's of his day, the suburbs were *the* place to live. Their situation afforded the 'maximum of comfort and enjoyment at a minimum of expense', combining nearby neighbours with the facilities of the town. To be near to a Botanic Garden, Loudon argued, was especially to be desired as it promised a source of year-round enjoyment. He felt that in industrial towns such as Manchester, householders were gardening in an artificial situation and that this needed special consideration when constructing a town garden. The possibilities of such a plot were endless as a small greenhouse and heated pits could be easily be added (especially once the glass tax had been repealed, in 1845).

By 1843, as the map of Ardwick Green shows, many changes were taking place. The encroaching terraced houses and industry threatened the suburban dream for the residents of the former fashionable inner suburbs.

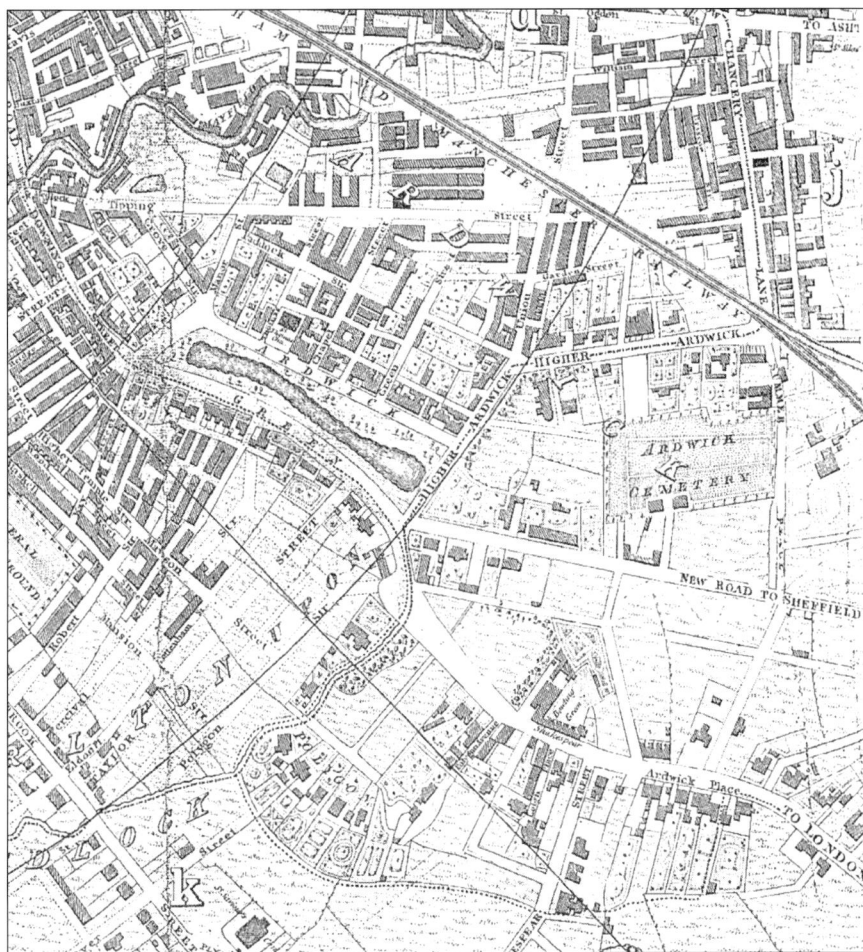

Ardwick Green and its environs, Manchester, 1843 (A Plan of Manchester and Salford with their vicinities, embracing every improvement, from actual survey, 1843, in *Pigot & Slater's General and Classified Directory*.
CHETHAM'S LIBRARY, MANCHESTER

As the nineteenth century progressed it is clear that the wealthier members of the Botanical Society, who could pursue their expensive botanical and horticultural interests as the founder members had done, chose to move further and further out to new Manchester suburbs, Didsbury or Chorlton-cum-Hardy, and beyond, all aided by a growing railway system. Suburbanisation was a phenomenon of Manchester's middle classes, as of other industrial cities, and this outward movement had an impact on the Botanical Society and the Manchester Botanic Garden. Residents of Manchester's new suburbs started their own local horticultural societies with exhibitions for members. The Cheadle Floral Society's advertisement in the *Gardener's Chronicle* for their 13th annual show in 1880, giving a date of 1867 for the Society's foundation, shows that many members of the Botanical Society had joined and were exhibiting in their own local societies to great success.[8] The Didsbury and Barlow Moor Flower Show in 1885 lists three winners who were also members of the Botanical Society: Messrs Brockbank (metal merchant), Morris (metal merchant) and Silkenstaat (cotton merchant).[ix]

Advertisements in gardening magazines of the time provide a wealth

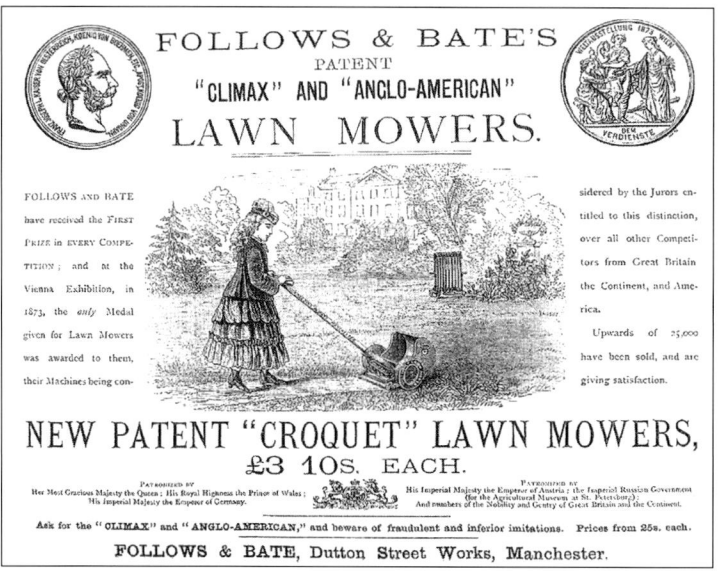

Follows & Bate's advertisement for lawnmowers. *Gardener's Chronicle,* 1870s, several dates.
ROYAL BOTANIC GARDEN, EDINBURGH

of information on products available and show how this industry was as widespread as it is today. The advert for the Manchester firm of Follows & Bate's portrays all that these new suburban villa gardens stood for. The house has its conservatory surrounded by a garden, a seat for leisure, pots for displaying ornamental plants and beds for trees, shrubs, and colourful bedding plants, together with the croquet set for recreation. This was a private place to be shared with family and friends. The villa gardens were now the suburban retreat from the industrial world, an individual's *hortus conclusus* made possible by the wealth created in the industrial society the owner sought to escape. The Manchester Botanic Garden had been such a private enclosed garden for the membership in 1831, and in 1888 it still fulfilled this purpose on the public stage at shows and exhibitions. By this late date however, members preferred their own private gardens for leisure and this offers an explanation for the lack of support for the Botanic Garden. As gardening activities and entertainment became centred on the home, the need for the Garden in the city decreased.

This was to be in the future. The concern of the members in 1827 was the establishment of the new Manchester Botanic Garden. First they had to find a suitable plot of land.

An Aside: the villa garden

Shirley Hibberd's requirements for a complete villa garden were given in his book *The Amateur's Flower Garden* first published in 1871. The front garden, planted with evergreens and bedding plants was bordered on the street by 'light iron railings'; this was the public face of the garden. In the private area essential elements included a screened vegetable garden, hedges and trees on the outer boundaries, a rockery with a pool stocked with fish and with a large vase in the centre as a fountain, a summer house, a greenhouse, flower beds, a rose garden, lawns with trees dotted in them, gardenesque beds, an American bed and a fernery. Joshua Major, another important landscape gardener with a national reputation, published *The Theory and Practice of Landscape Gardening* in 1852. He stated that the suburban villa, which might have land of an acre, but was exposed to public view if a belt of evergreen shrubs and low-growing ornamental trees was not planted around the boundaries. Major recommended that the beds were principally for good shrubs and the boundary was formed with shrubs and trees. He felt that it was inappropriate to try to imitate nature in a town garden and that owners should adopt a formal style of their choice.

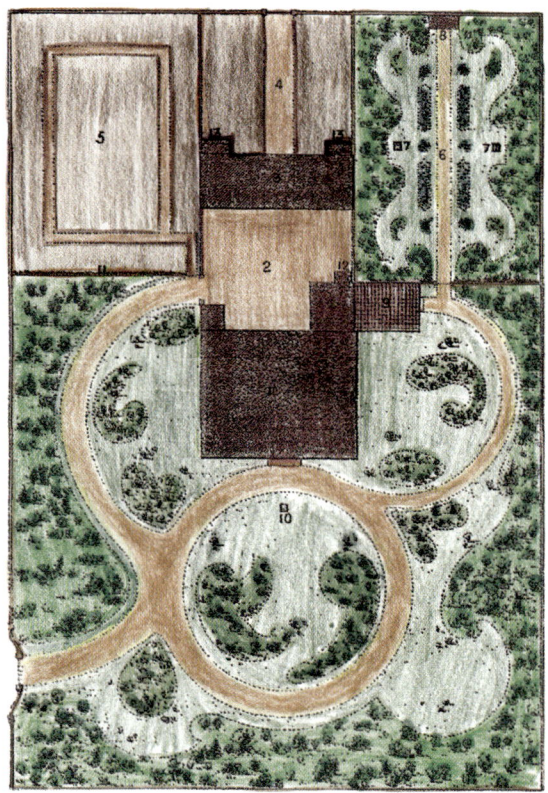

The Villa Garden: plan for a villa residence of 1 Acre. From J. Major, *The Theory and Practice of Landscape Gardening*, 188, pl. 2 (London, 1852).
Key: 1 House; 2 Yard; 3 Stables; 4 Clothes yard; 5 Vegetable garden; 6 Geometric flower bed; 7 Sculpture; 8 Covered seat; 9 Greenhouse; 10 Dial.
AUTHOR'S COLLECTION

Notes

1   *Report of the Select Committee on Public Walks*, Question 823, Answer, 59–66, (London, 1833). Manchester had a public walk in the Infirmary Gardens, Piccadilly.

2   *Mercury,* 9 October 1821.

3   J. Aikin, *A Description of the Country from Thirty to Forty Miles around Manchester,* 205 (London, 1795).

4   See C. W. Chalkin, *The Provincial Towns of Georgian England*, (London, 1974); Todd Longstaffe-Gowan in *London Town Gardens* comments that summerhouses and gardens were used as places of retreat *at some distance from the house* [my italics].

5   DDX 363, Archives, Caldwell's Nurseries, Knutsford, Cheshire County Archives, Chester.

6   P. Brown, MD, *A Catalogue of Very Curious Plants, Collected by the late Philip Brown, M.D. to be sold … at his Garden near Manchester* (Manchester, 1779).

7   *Transactions of the Horticultural Society of London* 1824, 53.

8   'Cheadle Floral Society', *Gardener's Chronicle* 4 September 1880, 313.

9   'Didsbury and Barlow Moor Flower Show', *Gardener's Chronicle,* 8 August 1885, 218.

CHAPTER THREE

A Botanic Garden for Manchester

Entrance to the
Manchester Botanic
Garden 1831. From *The
Mirror of Literature,
Amusement and
Instruction* 536, Vol. 19,
Saturday 3 March 1832,
129.

AUTHOR'S COLLECTION

Once the Council had decided to proceed their first priority was to find a site.
The second was to be the design for the garden. These were both achieved within
the next two years. First they concentrated on the land and, on 11 October 1827,
the *Chronicle* carried the following advertisement:

> The committee of management of the Manchester Botanical and Horticultural
> Society require a quantity of from six to ten Cheshire acres. The principal points in
> making this selection, besides the *nature of the soil and a proper supply of water,* will be
> as much as possible to avoid those nuisances of a manufacturing district prejudicial
> to vegetation – to secure an accessible distance and the advantage of a good road. It
> is conceived the site should be within three miles of the Manchester Exchange.

The closing date for submissions was 1 December 1827 and speculation as to
the site must have been widespread. The Council hoped that landed proprietors
would furnish details of plots available as they assured them improvements
would result if the Garden adjoined their property.[1]

The Council, aware of how critical ease of access to the Garden would be in
attracting members, was also supporting a proposed new turnpike from Old
Trafford to Stretford Road, Manchester.[2] The local aristocracy supported the
road and, by 26 April 1828, the *Guardian,* keen to point out the advantages
of the thoroughfare, claimed it would add value to the surrounding land and
would be more agreeable to the subscribers than 'the circuitous and crowded

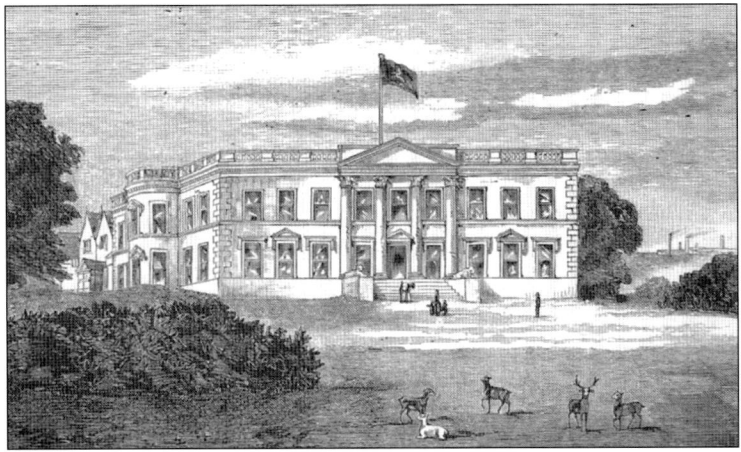

Trafford Hall. The gates
to the Botanic Garden
were directly opposite
the gates to Trafford
Hall on the Chester
Road. From A. Rimmer,
*Summer Rambles around
Manchester* (Manchester,
1887).

route through Hulme'. Importantly, the road allowed subscribers much easier access to the Garden by reducing the travelling distance from the centre of the town to less than a mile. In addition, it would be perceived as safer as they would avoid close contact with the working classes *en route,* which had raised the spectre of disease, unrest, and riots.

On 8 March 1828 the *Guardian* reported that 12 acres of land had been purchased from T. J. Trafford of Trafford Park, 'beyond the Hulme Toll Bar'. On 13 March the *Courier* stated that this report was wrong, as it was not 'within the province of the Committee themselves to conclude a purchase', claiming that a decision had not been reached and that several plots were being considered, though adding that the land offered by Trafford was 'on such terms as to guarantee selection'.[3] In a final flourish the *Courier* announced that they had just discovered that 'an individual who is both practically and scientifically acquainted with the requisites for a public Botanic Garden' was going to inspect the plots under consideration. The printed report sent to members on 21 April 1828 confirmed that an outside arbiter had been appointed. The Council had thought it sensible to obtain the views of someone trusted by the subscribers to justify their choice of the site.[4]

The man chosen was John Shepherd, Curator of the Liverpool Botanic Garden, who was well known to the Manchester gardening elite as he had formerly been gardener to John Leigh Philips, a Manchester antiquarian. Philips, a close friend of William Roscoe, had donated several rare plants to the infant Liverpool botanic garden, and was well known in botanical circles. By 1827 Shepherd was highly regarded within the network of subscription botanic gardens, a fact well known in Manchester.[5]

After proclaiming that they had not confined their views to any particular district and investigated all their natural advantages, the Council reported that they and Shepherd were in agreement on the site at Old Trafford.[6] The Council did allude to a possible argument about which site should have been chosen as 'motives of economy may prevail over any advantages either of soil or situation', but this

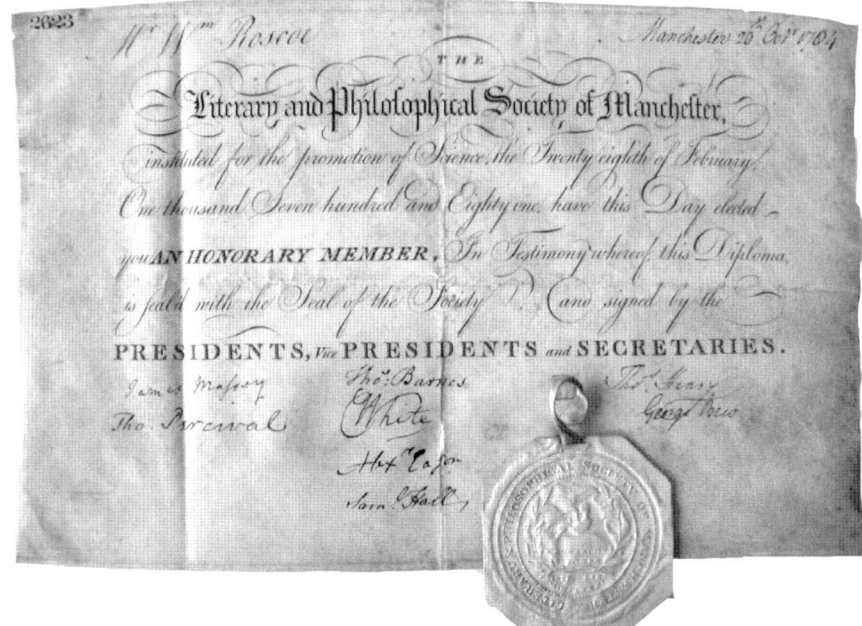

Roscoe's Honorary
Membership Certificate,
The Manchester Literary
and Philosophical
Society, 1781.

seems to have been resolved by Shepherd's recommendation of Old Trafford.
This description of events seems to contradict the local myth that John Dalton
determined the location of the Garden on meteorological grounds. Benjamin
Love gave one version in 1839, claiming that Dalton, based on his 40 years of
weather studies, declared that Old Trafford was free from smoke most of the
year.[7] A different version appeared in June 1896 in the obituary of Bruce Findlay,
Curator of the Manchester Botanic Garden, reporting that 'Dr Dalton went all
around the suburbs of the town, testing the comparative cleanliness of the leaves'
and that the purest air was found in Old Trafford.[8] No evidence has been found
to confirm this appealing story. The recommendation to purchase the land at
Old Trafford was accepted unanimously by the general meeting on 21 April 1828.[9]

The Council arranged for a site survey from a Manchester surveyor,
W. Johnson, to establish the nature of the soils. This met all the Council's
requirements. The annotated plan showed the land falling away south from
the turnpike road to a point 17 ft (*c.*5.2 m) below. It consisted of three distinct
types of soil: the largest area, nearest to the road, was 'Rich fine deep soil';
the second area, to the bottom right, was 'Fine soil but of somewhat inferior
quality'; and the final area, bottom left, which also contained pits, was 'Strong
clay to supply bricks for the wall and buildings'. The Council had found an
attractive site which was described by Joseph Paxton and J. Harrison, editors
of the *Horticultural Register,* on their visit to the Garden in 1831, saying, 'The
situation of the garden is admirably chosen for the purpose ... having a full
command of an extensive tract of rich level country, terminating with a view
of the Derbyshire and Staffordshire hills'.[10]

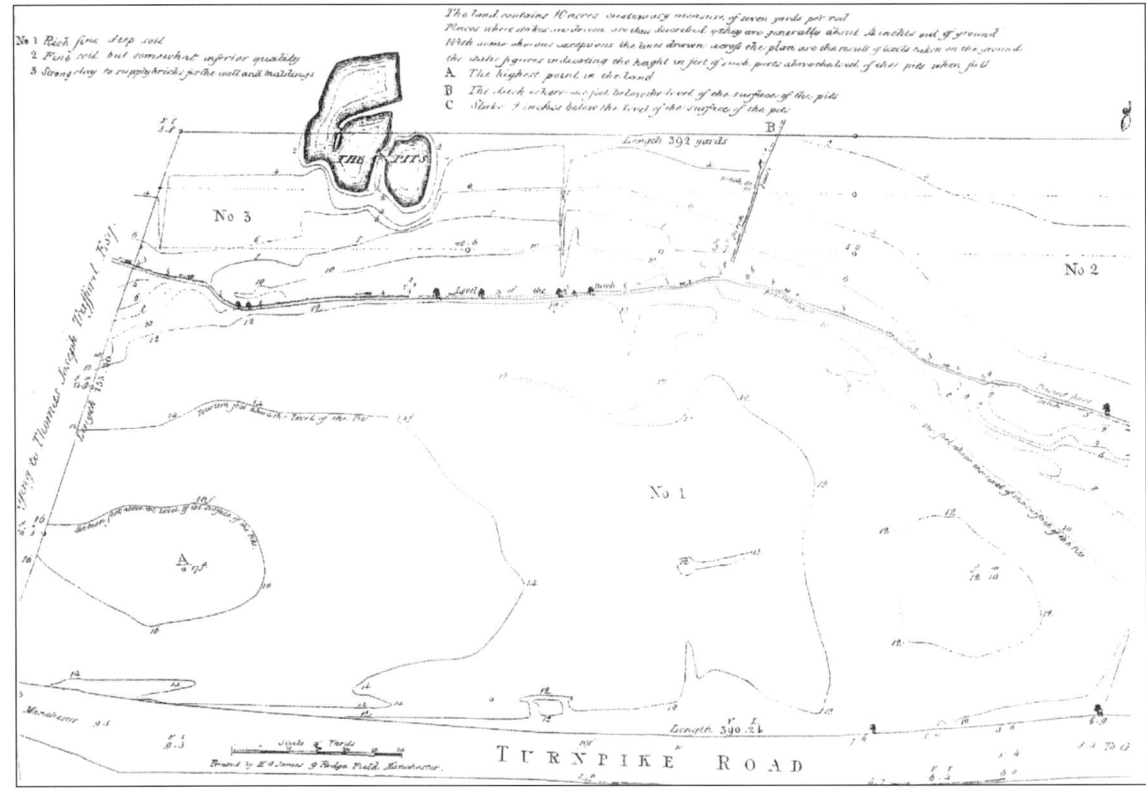

Johnson's Land Survey. (Plan of Survey, EGR 4/2/10/20/8). See also 'Title Deed', 3 February 1829, MBH 1/1. William Johnson was a land surveyor and valuer, 33 Brown Street, Manchester. From *Pigot's Directory* (Manchester, 1832).

REPRODUCED BY COURTESY OF THE UNIVERSITY LIBRARIAN AND DIRECTOR, THE JOHN RYLANDS UNIVERSITY LIBRARY, THE UNIVERSITY OF MANCHESTER

The Society had been offered the land on very attractive terms as T. J. Trafford had let them decide the purchase price; they had set it as the acreage price for local farmland. The final price was £192 together with a 99 year, perpetually renewable, lease and an annual ground rent of £120. This was indeed a generous bargain for the Society as the Birmingham Horticultural Society was paying more than double the sum to Lord Calthorpe for land on his Edgbaston estate for their Botanic Garden, on exactly the same acreage of ground for a 60 years' lease.[11] To be fair to Lord Calthorpe the land in Birmingham was part of an already-developed estate of more value per acre, rather than farmland as in Manchester. The site in Birmingham was on the Calthorpe Estate where the policy, with its strict covenants, was to develop large detached villas in their own grounds attracting the Birmingham merchant and professional classes.[12] A few leases had been granted to local institutions and the application for the Botanic Garden succeeded on the grounds that it would 'greatly promote the interest of the estate' and, being in the central area, was easily accessible and so regarded as an amenity for all.[13] It seems that Trafford hoped the Botanic Garden would be an attraction to Old Trafford and he could develop this part of his estate as expensive housing.

The conveyance of the land was drawn up on 3 February 1829.[14] This shows that Trafford, like Lord Calthorpe in Birmingham, was interested in maintaining the amenity of the site itself, together with the value of his retained

View of Castleton, Derbyshire. From J. Aikin, *A Description of the Country from Thirty to Forty Miles round Manchester*, 465 (London, 1795).

land. Within two years of the date of the conveyance, the Society had to erect at least one substantial dwelling house with a yearly rental value of at least £30 confirming that Trafford clearly intended Old Trafford to become a high-value residential suburb. Covenants listed prohibitions intended to maintain the local environment. The Society was prohibited from allowing drinking houses, steam or fire engines, vitriol work, or other manufacturing trades or businesses on the site. Ironically, the later industrialisation of the retained Trafford Estate was to be one of the factors that damaged the viability of the Garden. Though the Society had obtained the agreement of its members to purchase the land in April 1828, it was February 1829 before the final deeds were signed.[15]

In the meantime the Council had not been idle. On 30 August 1828 it advertised a competition for designs for the Botanic Garden and explained that 'Lithographic drawings of the grounds are to be had, with every additional information' (no copy was found).[16]

Although J. C. Loudon, the arbiter of all things horticultural at the time, claimed a botanic garden was 'scientific above all else', a study of some of the design elements suggests the possibility of a different interpretation. For example, rockwork could be seen botanically as a habitat for alpines, ferns, and mosses, yet Paxton and Harrison's description of the Manchester rockwork makes it clear that it was admired on its own account for its picturesque qualities as it was '... in the best possible taste; we might fancy ourselves roaming in some of the lovely valleys of Derbyshire, with the rivers Wye, Derwent, or Dove, rippling at our feet'.[17] This was one of the elements in the Garden that could be attributed to the Romantic Movement. The founders in their *Address* had anticipated this apparent contradiction, between the scientific and the fashionable pleasure ground, when they claimed that the Garden would appeal to both tastes. Loudon, too, was aware of this dichotomy and when asked to design the Birmingham Botanic Garden incorporated their request 'to combine a scientific with an ornamental garden'.[18] Whether the function of the Garden was educational or recreational, it will become clear, was to prove an issue for the Manchester Society.

The advertisement that appeared in the *Chronicle* on 30 August 1828 made it clear that there were to be prizes for the best designs. The Council offered 50 guineas, 30 guineas, and 20 guineas and if the plans were delivered on or before 29 September, they would be shown to the public at the Society's forthcoming flower festival. The Council did not require specifications of the work, simply drawings of the 'elevations' and a plan of the garden. To guarantee anonymity, each plan had to be identified by a number or motto on the sealed envelope containing the name of the contestant. Unfortunately none of the plans has survived.

> To Architects, Designers, *etc.*
> The Committee of the Manchester Botanical and Horticultural Society being now to proceed to the formation of a PUBLIC GARDEN are desirous of obtaining a plan adapted to the nature of their Institution and propose a liberal premium for the best design that may be produced.[19]

It is instructive to see the use of the term Public Garden in this advertisement. The use of this term is not as we understand it today. Though in theory the *Address* had appealed to all classes the very nature of the enterprise meant that only those who could afford the subscription could support the garden. In reality this meant the local aristocracy, gentry, and upper middle classes. The latter were an expanding group within the town as the industrial revolution had brought wealth to merchants, manufacturers and associated professions. The 'Public', who would enjoy the facilities, were those who subscribed and this guaranteed them an exclusive garden. Again, as with the Garden's function, it will become clear that the concept of exclusivity is central to the history of the Manchester Botanic Garden.

At a meeting of these wealthy subscribers on 26 November the Chairman, Hugh Hornby Birley, announced that six entries had been received and declared the winners were:

> The First to the plans bearing the motto 'Hope'. The second to the plans bearing the motto '2'. The third to the plans bearing the motto '*Ut flos in sipis secretis nacitur hortis*'. That the author of the plans, accompanied by the motto, 'K828K' be requested to accept a sum of 10 guineas on condition ... his plans remain with the Society.[20]

First prize went to Mr J. B. Watson, Portman Square, London; second to Messrs Royle and Irwin, Manchester, and third to Mr Jenkins, Red Lion Square, London.[21] The addresses of the winners make it clear that the competition had generated both local and national interest. Royle and Irwin were architects, surveyors, and agents with offices at 23 Charlotte Street, Manchester. Finally the Council was empowered by the members 'to decide upon the final plan, form contracts, and establish the garden'. The Council still had a dilemma, as a report in the *Courier* on 29 November makes clear. Though prizes had been awarded to the best three, the Council could not decide which to recommend: perhaps, it was noted, suggestions could be selected from each one. In effect the Council did not use any of the winning designs and the first Curator, William Mowbray, planned the Garden. A fact acknowledged by Loudon after his visit in 1831: 'it is but justice to Mr Mowbray, to state that the present plan, which is almost entirely his own, is greatly superior to all the different plans which were

An Aside: William Mowbray, Cuator 1828–1833

'… he has left behind him a character crowned with great respect.'[1]

William Mowbray, Manchester Botanic Garden's first curator, was born in Hitchen, Hertfordshire in 1792 where his father was a gardener and seedsman. As a journeyman gardener, he was employed by the Comte de Vandes at his private botanic garden in Bayswater, London.[2] Before coming to Manchester, his obituary reveals, he was gardener for 11 years to the Earl of Mount-Norris, Arley Hall, Bewdley.[3] While there, Mowbray contributed to the *Transactions of the Horticultural Society of London*. The first, read on 3 December 1822, concerned the cultivation of *Mesembryanthemums*.[4] The second, published in 1825, described his successful trials obtaining fruit from species of *Passiflora*.[5] An article by him appeared in *The Gardener's Magazine* May 1832, 'On the raising of new varieties of the Tree Paeony, (*Paeonia Moutan Banksia*)'.[6] Mowbray was a skilled horticulturist and also the designer of the Manchester Garden. Loudon said his design was 'vastly superior' to all other plans submitted to the Society.[7] His last months at the Garden are not recorded in the Society's minutes. His health was giving concern in March 1832, when the minutes record that he apologised for being unable to go on a visit to the south of England. Mowbray died on 10 July at Hitchen, Hertfordshire, where he retired because of his protracted ill-health. He was in his forties and suffering from consumption (tuberculosis). The author of his obituary attributed his death to the strain of constructing the Manchester Garden and claimed that the Manchester Botanic Garden was 'a monument to his fame, as a scientific and practical gardener.'

Passiflora caerula, introduced from South America, especially Brazil, in 1699.
AUTHOR'S COLLECTION

1 Obituary: William Mowbray, *Horticultural Register*, 4 August 1832, 670. All information is from this source unless otherwise stated.
2 *Gardener's Magazine*, August 1831, 41.
3 *Gardener's Magazine*, August 1832, 670.
4 Letter, 30 November 1822, XXXIII, *Transactions of the Horticultural Society of London* 4, 274–5.
5 Letter, 29 October 1824, *Transactions of the Horticultural Society of London* 5, 95–6.
6 Article XII, Letter, 10 April 1832, *Gardener's Magazine*, May 1832, 500–1.
7 *Gardener's Magazine*, August 1831, 414.

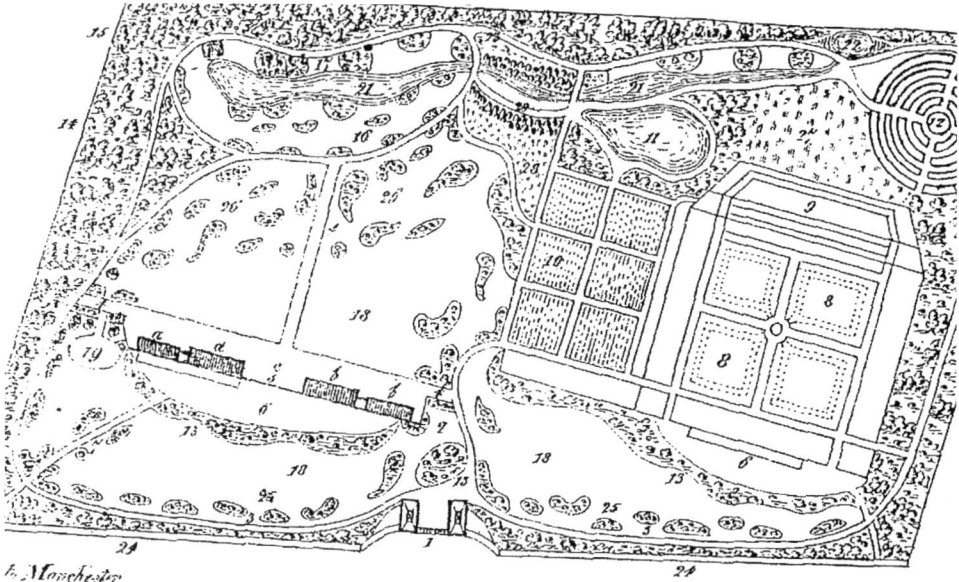

h. Manchester

Plan of Manchester
Botanic Garden.

HORTICULTURAL REGISTER 1
SEPTEMBER 1831
© ROYAL BOTANIC GARDEN,
EDINBURGH

sent in for competition'.[22] The Council had appointed Mowbray as Curator in 1828, certainly before the land had been purchased and possibly at the time of the design competition.

The ground plan of the Garden shows the elements contained within it until 1848 when major changes took place. The plan is orientated looking south from the Chester Road, *i.e.* south is to the top of the map. The numbered references relate to those on the plan. The entrance (1) shows there were two lodges.[23] Mowbray included a belt of trees, the arboretum, immediately within three sides of the surrounding wall (15). On the fourth side, the wall was to be used for climbers and under-planted with annuals (24). The lower ground had needed draining, as a stream flowed the length of the garden, and he had capitalised on this to provide water for the aquarium and bog garden (21; 22). A stove-house and greenhouse were both complete by 1831 (a), and hothouses (b) were then in the process of construction. The central house, the conservatory (c), had not been started because of lack of finances at the time the article was written. There was a fruit and vegetable garden with heated walls, which contained heated pits for pineapples, an experimental garden and 2 acres of standard fruit trees (8). An outdoor bed for greenhouse and alpine plants during the summer months was sited next to the greenhouse and stove, for convenience of transport (19). There were eight separate compartments, each for a collection of specified herbaceous plants, laid out scientifically according to the Linnaean system (10); medicinal plants were planted separately (9). The rockwork was intended to be extensive, with caves and vaults, tunnels and pendant walls (20), and Mowbray described how a 'Dell', part of the rockwork, was formed when excavating sand for the bricklayers. The second 'Dell', containing the majority of the rockwork, was excavated when the lake was constructed.

In addition to the Council's specifications, Mowbray had incorporated a mound (2) and rosary (12). The rosary was of an earlier geometric form, allowing visitors to stroll around the paths and view the plants from all sides surrounded by their scent. The mount (or mound) had been a feature of English Tudor gardens that survived into the eighteenth century and the gardens at Dunham Massey contains a noted example. The mound was planted with evergreens which would offer shade when they matured and, placed near the entrance, took into account the fall on the land and allowed the subscribers to enjoy the view over the Garden to the hills beyond. A different purpose for the mound was given by a correspondent of *The Gardener's Chronicle* in 1857: 'Opposite the entrance is a large mound planted with shrubs and evergreens, the principal object being to prevent a direct view into the gardens from the road, and also to break the force of the north winds'.[24] An examination of the plan shows a bed of trees and shrubs directly opposite the entrance (13), adjacent to the mound, which could have been planted as a windbreak and screen. As it was common practice to plant in slightly raised beds, and given that the trees would by then have been 26 years old, it seems probable that it was this matured feature that was being described as the mound in 1857. Or perhaps the mound always fulfilled a dual purpose enhanced by the adjacent trees. Finally, gravel paths wound round the site (3, 4), and beds containing flowers, shrubs and trees were placed here and there in the lawns (25, 26).

By 1828 the Council had the plan for the Garden and in February 1829 the land had been purchased. Construction could begin at last.

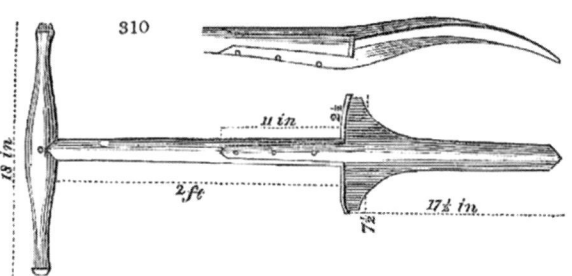

The Perforator. From J. C. Loudon, *An Encyclopaedia of Gardening New Edition* (London, 1834).
AUTHOR'S COLLECTION

Notes

1   *Guardian*, 26 April 1828.

2   Report on a meeting of Trustees of the Turnpike Road, Letter, Hugo Worthington (Agent to the Earl of Stamford and Warrington at Dunham Massey) to the Earl, 14 June 1830, G/2/2/9/2, GB-2184-Grey, Earls of Stamford and Warrington, Enville, Worcestershire; Report of a meeting of the Committee for the new branch road from Old Trafford to Manchester, Letter, Worthington to the Earl, 6 July 1830, G/2/2/9/3, *ibid*. This includes names of the landowners affected.

3   *Courier,* 13 March 1828.

4   *Report of the Committee of Management* (Manchester, 1828), EGR 4/2/10/20/6.

5   Liverpool City Council, *Wavertree Botanic Gardens,'A hidden gem in the crown of Liverpool', a feasibility study for their regeneration* (Liverpool, 1989).

6   *Report,* EGR 4/2/10/20/6.

7   B. Love, *Manchester as it is,* 121 (Manchester, 1839).

8   'Memorial Notices Bruce Findlay', *Guardian,* 17 June 1896.

9   *Guardian*, 26 April 1828.

10   'Description of the Botanic and Horticultural Garden Manchester', *Horticultural Register,* 1 September 1831, 109.

11   J. C. Loudon, 'General Results of a Gardening Tour', *Gardener's Magazine,* 31 August 1831, 415

12   See D. Cannadine, *Lords and Landlords: the aristocracy and the towns 1774–1967,* 157 (Leicester, 1980)

13   Loudon, 'General Results of a Gardening Tour', *ibid*.

14   'Title Deed', MBH 1/1. This document also confirms Trafford's reason for selling the land as it states 'the plot … would *add considerably to the value of the adjoining lands and property of the said T. J. Trafford* [my italics]'.

15   *Counterpart Demise in Trust,* MBH 1/3.

16   *Chronicle,* 30 August 1828.

17   *Horticultural Register,* 1 September 1831, 109.

18   Quoted in Ballard, *An Oasis,* 18.

19   Design specifications and competition details, Manchester Botanical and Horticultural Society, Letter, 1 September 1828, EGR 4/2/10/20/7. As this was sent to the Earl of Stamford and Warrington it seems probable that copies of the design specifications were sent to all Patrons and other Botanic Gardens.

20   *Chronicle,* 29 November 1828.

21   *Manchester and Salford Directory and Memorandum Book for 1828* (Manchester, 1828).

22   *Gardener's Magazine*, August 1831, 414.

23   Shown clearly on the remaining façade, though the original was modified in 1907 by The White City Co Ltd.

24   *Gardener's Chronicle,* 9 May 1857, 327.

CHAPTER FOUR

The Membership
'Individuals of the highest respectability'[1]

An Uncharitable view of a Charity Ball from *Punch:* 'the large number of fancy dresses (which) arrived in Manchester to be hired out which were sent by firms in London, and no doubt from the continent … the day before I was able to hire a dress, a Turkish sailor, for five shillings.' *Punch* vol. 4. London, 1843, 87. Text, J. T. Slugg, *Reminiscences of Manchester Fifty Years Ago*, 305–6 (Manchester, 1881).

BOTH PORTICO LIBRARY, MANCHESTER

On 30 September 1828, the first Manchester Musical Festival opened and as a grand finale on 4 October, a fancy dress ball was held at the Manchester Assembly rooms on Mosley Street.[2] This social event was attended by many of the members of the newly formed Manchester Botanical Society, for example: Rev. John Clowes went as himself, E. J. Loyd in the full dress uniform of the Cheshire Yeomanry Cavalry, Peter Ewart in court dress, and Samuel Brooks the banker was a steward in 'elegant court dress'. The local aristocracy were present as well as established tradesmen of the town, for example Mr Blackberd, a seedsman and active member of the Floral Society. At the ball, social activities in the town such as the founding of the Botanic Garden must have been a lively topic of discussion and perhaps members were recruited in this easy atmosphere.

When the Manchester Botanical and Horticultural Society was founded in 1827, who were to be the subscribers and what were their expectations? The establishment of the botanic garden was not only a response to a national trend for subscription botanic gardens but also a draw for those with an interest in scientific botany, gardening, and horticulture, who already had other societies in the town to choose from. The membership lists of these societies shows members were drawn from the upper middle class in terms of both religion and politics and often had aristocratic connections as either members or patrons.[3] These Societies, and others nationally, had links to the burgeoning scientific developments in the natural world. Though few of these early societies specifically targeted gardeners, they had common organisational features that would be later adopted by the Botanical Society. For example, the Societies discussed below had links to scientific developments in the natural world and their membership lists show that some members were amongst those founding the Botanical Society.

The Manchester Agricultural Society, founded in 1767, did not consider horticulture to be within its ambit, though it was interested in the application of science to agriculture. The Society had aristocratic patrons and offered prizes for competitions.[4] There were also corresponding members who were 'the most eminent characters in Europe'. By 1804, 'it was composed of one hundred and sixty-seven subscribers, of the greatest respectability, in rank, fortune and talents.' Manchester's middle class elite was sympathetic to its aims, as shown by the number of members listed as living in the town. It was this town elite who would later establish the Botanical Society.

One society connected with the sciences of the natural world was The Manchester Society for the Promotion of Natural History founded in June 1822.[5] The report of the annual meeting on 14 January 1830 showed that they, too, had aristocratic patrons: The Rt Honourable the Earl of Wilton and Sir Oswald Mosley, Bt., both later Patrons of the Botanical Society. The Committee was composed of 21 members of the upper echelons of the town; eight became members of the Council of the Botanical Society including its President, Edward Holme, MD.[6] In 1839 Benjamin Love wrote:

> The museum, in Peter-street, was established in the year 1821, its foundation being a collection of British and foreign insects, collected by the late Lee Phillips, Esq. … At the present time, the ornithological collection stands the first in the provinces of Britain, if not of Europe. The other departments are not so richly supplied. … but with such funds as the society possesses, it may, with good management, be made to rival some of the first metropolitan institutions.[7]

The Society's Museum and Library received 69 donations in 1830, which came from sources as distant as Tunis, the Isle of Uist, and North Wales, a practice common amongst subscription botanic gardens.

The Manchester society devoted to gardening was the Floral and Horticultural Society founded in 1824, and advertisements appearing in the local press demonstrate that exhibitions and competitions were held regularly throughout the year and prizes were awarded for florists' flowers, stove, greenhouse, and herbaceous plants, shrubs, vegetables, and fruit.[8] Membership was by subscription and entitled subscribers to free entry with non-subscribers paying 1 guinea.[9] Typical of the meetings was the description in August 1829 as 'a numerous and fashionable assemblage of company.'[10] J. C. Loudon, visiting one

The Manchester Exchange. Shows were held in here in the Manchester Exchange dining room and the Town Hall. From S. Austin, J. Harwood G. and C. Pine, *Lancashire Illustrated* (London, 1832).

of their shows in 1831, claimed he had never seen so many excellent specimens of hothouse, greenhouse, and hardy plants assembled together and that 'these competitions … were the mainstay of the exotic nurseries in London'. The aggregate prize money was between £500 and £600 and he claimed, 'so eager are the wealthy to obtain these [prizes]' that they ordered the latest expensive plants from London. He explained that the members read *The Botanical Magazine* and the *Botanical Register* each month and, to forestall their rivals, immediately purchased the new introductions in case stock was low. Only the wealthy could afford this style of gardening.

Indeed the committee members of the Floral Society were instrumental in the foundation of the Botanical Society. In August 1827, the same month as *An Address to the Inhabitants of Manchester and the Neighbourhood on the formation of a Botanical and Horticultural Garden* was published, a letter appeared in the *Manchester Guardian* commenting on the proposed Garden.[11] The anonymous author, 'A Florist', claimed that '*An Address*' was being circulated 'amongst the members of the Floral Society, who first brought the subject before the public', and stated that many of Manchester's distinguished and influential citizens had sanctioned the new enterprise and intended to patronise the endeavour. Certainly the two societies had similarities: membership by subscription, free entry to shows, and wealthy members able to grow specialist plants that required heated conditions.[12] The report of the annual meeting of the Floral Society for 1831 gave the names of the Committee, of whom some had been founding members of the Botanical Society.[13] All but one of the Officers of the Floral Society became members of the Council of the Botanical Society. A letter 21 July 1832 in the *Manchester Times* confirmed the connection between the two Societies claiming that 'I was informed that this Society [Botanical and Horticultural] was but a *scion* from the original; and their views were lofty and aristocratic'.[14] The Manchester Floral and Horticultural Society foundation in 1824 was perhaps a response to the failure to found a botanic garden in 1822. After three successful years, it seems probable that the Floral Society again mounted a campaign for a Manchester Botanic Garden in 1827.

A subscription to the Botanical Society gave access to an exclusive club. Exclusivity was to be a defining feature of the Manchester Botanical and Horticultural Society throughout its history. To achieve this, access to membership had to be controlled by means that ensured that only those who were socially acceptable were admitted. Two measures were available to the Council; firstly, setting the fees for membership at a level that excluded all but the well-off and secondly, peer approval of candidates. All classes of members, hereditary, life, and annual, had to be proposed by a subscriber and their applications approved by the hereditary members by ballot. Both these devices were applied rigorously. Even in the 1850s, when the level of fees had to be modified to attract more members, continuing to ballot for final approval meant social exclusivity could be maintained. It was ironic that, in 1898 when the Society's garden was no longer advantageously situated and under such enormous financial pressures that it had been leased out for public attractions, a member of the Council could still assert that: 'general use of the gardens by the "masses" has led to the falling off of the "classes"'.[15]

Patronage by the aristocracy was another facet of exclusivity. Patrons of the Society had been officially confirmed in 1829 as The Earls of Stamford and Warrington (Dunham Massey) and Wilton (Heaton House), Lord Suffield (Middleton, Lancashire), and T. J. Trafford (Trafford Hall).[16] This form of Patronage was common practice within the subscription botanic garden movement; the 6th Earl of Stamford and Warrington was also both Patron and member of the Birmingham Botanical Society.[17] This was essential for the Botanical Society as it is clear that such patronage was essential to attract members. For the Society the ultimate social accolade was granted in 1876 when Queen Victoria became Patron and competed at the Society's exhibitions and the sobriquet Royal was added to the title, The Royal Manchester Botanical and Horticultural Society.

The Botanical Society Council of 1827 was also a reflection of the changing town council. The Unitarians were challenging the old Tory leadership of the town. They had emerged as cultural and civic leaders, wealthy and respectable, as well as liberal in politics and theology. On the Society's Council there were prominent Unitarians, Richard Potter (later MP for Wigan) and the bankers Benjamin and Benjamin Arthur Heywood and staunch Tories, Hugh Hornby Birley, major of the Manchester Yeomanry Cavalry at Peterloo in 1819 and first president of the Manchester Chamber of Commerce in 1822, and Gilbert Winter, wine merchant and Boroughreeve of Manchester in 1823–4. Other

Introduced from Mexico in 1809, *Ardisia crenulata* has reddish-violet panicles of flowers and long-lasting bright coral berries. from *The Illustrated Dictionary of Gardening* 1, 10 (London, n.d.)

non-conformist members included Robert Hyde Greg, of Quarry Bank Mill, Styal, and MP for Manchester 1839, Thomas Heywood, Boroughreeve Salford, 1826, and the scientist, William Henry.[18] Though in town politics these were two mutually antagonistic groups, we see in the Botanic Society an example of how members of different political and religious groupings in Manchester co-operated in social enterprises.[19]

Unusually for Societies of the time, women were subscribers to the Manchester Botanical Society from the beginning, appearing in the first surviving membership list (1833) where they formed 3% of the total membership (Appendix 1). The subscription for hereditary, life, and annual members always included the member's immediate family, so women of the family were allowed entry to the garden as of right; the rule excluding sons over 18 did not apply to daughters. Even when there was a hereditary male member living at the same address, women appear on the membership lists in their own right. Under the normal circumstances, married women could not own property and were excluded from societies that owned their own premises in the name of the members.[20] There is an explanation for the anomaly. When the land had been purchased there was a clause stipulating that if the Society was to sell the Garden all proceeds had to be spent on botanical and horticultural endeavours. This meant in effect that members did not have ownership of the land; it was vested in trustees. As botany was a subject approved for study by women, and the legal problem of ownership irrelevant, they were specifically targeted as members. 'A Florist', in his letter of 2 August 1827, confirms this when he appealed to Manchester's ladies when seeking support for the newly established botanic garden. There, he argued, they could study botany and take pleasurable walks.[21] He felt that ladies had a sacred obligation to help, as their influence on their husbands would determine the character of the town:

> And shall it be said the Ladies of Manchester are insensible to the attractions of a Botanic Garden? Shall they who receive the freeman's homage, be deaf to the appeals of the beauties of Nature? If this interesting science can excite no sympathies in

The New Jerusalem Church, Manchester. The Swedenborgian Church on Peter St, Deansgate showing smoking industrial chimneys near the city centre by 1832. From S. Austin, J. Harwood G. and C. Pine, *Lancashire Illustrated* (London, 1832).
AUTHOR'S COLLECTION

their bosoms, the labours of our townsmen have been lost, and the wealth of our forefathers expended in vain.

On 17 March 1831, Mrs Diana Beaumont of Bretton Hall, Yorkshire, who had donated plants and been consulted on glasshouses, became the first female Honorary Member of the Society.[22] The second, 2 November 1833, was Mrs Hobson of Hope Hall, Eccles Road – she appeared in the 1833 membership list of the Botanical Society as did her husband, Edward. Mrs Hobson also appeared regularly as a prizewinner in the competitions of both Societies, always exhibiting in the sections for stove plants, the most expensive form of gardening. She won the premier stove plant class with a specimen of *Ardisia crenulata* at the Botanical Society's first exhibition in June 1831.

At the Floral Society's second exhibition in May 1832, she repeated her success with the same plant.[23] Was this a further connection between the two Societies? Was she a leading member of the Floral Society, who helped establish the Botanical Society and generously donated plants for the stoves and greenhouses of the infant Manchester Botanic Garden? In gratitude the Council awarded her the same honorary membership as it did to aristocratic donors.

What drew these diverse groups of subscribers together was a shared interest in science and botany coupled with the privileges of a private garden within an industrial city. All were important citizens establishing the botanic garden as both an ornament to Manchester and as a reflection of their own enlightened taste. The Garden was to be their *hortus conclusus* within the burgeoning industrial town.

An Aside: comparison of Committees

Botanical and Horticultural Society
4 August, 1827 (*Mercury*)

Patrons:
Earl of Stamford and Warrington
Earl of Wilton
Lord Suffield
T. J. Trafford

President: Edward J. Loyd Barrister

Vice-Presidents:
Hugh Hornby Birley Manufacturer
Rev. John Clowes Landowner
R. A. Heywood Banker

Treasurer: Richard Potter Manufacturer

Hon. Secretary:
Rev. Hordern Chetham's Librarian

Committee:
John Barton Surgeon
R. W. Barton Merchant
James Benson Merchant
Thomas Boothman Jn. Merchant
William Bow Gentleman
Samuel R. Brookes Merchant
Thomas Edmund Buckley Merchant
James Darbishire Manufacturer
Thomas Fleming Manufacturer
Thomas Glover Merchant
Benjamin Heywood Banker
Thomas Heywood Banker
William Charles Henry MD Physician
Robert Hindley Brewer
George Hole Calico Printer
John Hull MD Physician
Thomas Markland Merchant
John Milner Marris Merchant
John Moore
Shakespeare Phillips Manufacturer
Edwin William Sargeant Attorney
Thomas Sharpe Merchant
Christopher Todd Manufacturer
Thomas Turner Surgeon
Charles Walker Calico Printer
Gilbert Winter Merchant
G. W. Wood Banker

Floral and Horticultural Society
17 February, 1827 (*Guardian*)

Patrons:
Earl of Wilton*
T. J. Trafford*

President: Rev. John Clowes A. M.*

Vice-Presidents:
R. W. Barton* Merchant
T. Heywood Esq.* Banker
William Hulton Esq.
Richard Potter Esq.* Manufacturer

Treasurer: Thomas Boothman*

Hon. Secretary:
James Benson* Merchant
John Milner Marris* Merchant

Committee:
Mr. James Walker Ashton
Mr. J. Benson Jn.* Merchant
Mr. T. Boothman Jn.* Merchant
Mr. J. W. Boothman
Mr. William Bow* Gentleman
Mr. James Darbishire* Merchant
Mr. E. R. Fletcher
Mr. James Glover
Mr. Thomas Hadfield
Mr. H. Hadfield
Mr. Thomas Knight
Mr. Samuel Lees Merchant
Mr. John Milner Marris* Merchant
Mr. Edwin Sergeant* Attorney
Mr. William Taylor
Mr. Thackery
Mr. William Thomas
Mr. Christopher Todd* Manufacturer
Mr. John Wakefield

Notes

1   Quoted in 'The Manchester Botanical and Horticultural Society', *Chronicle,* 27 September 1831.

2   *An Account of the Manchester Musical Festival 1828 containing the names of the Patrons and Committee; A report of the Oratorios and concerts and a list of the Principal Vocal and Instrumental Performers with a description of the characters who attended the Fancy Dress Ball* (Manchester, 1828), R780.64 mc646 (Henry Watson Music Library, Manchester Central Library, Manchester).

3   A. J. Kidd and K. W. Roberts (eds), *City, Class and Culture,* 12 (Manchester, 1985).

4   J. Aston, *The Manchester Guide,* 229–30 (Manchester, 1804).

5   W. E. A. Axon, *The Annals of Manchester,* 165 (London, 1886).

6   *Guardian,* 16 January 1830.

7   B. Love, *Manchester as it is,* 124 (Manchester, 1839).

8   W. E. A. Axon, *The Annals of Manchester,* 168 (London, 1886). Manchester newspapers had reported on Florists' Societies since the eighteenth century. For information on Florists' Societies see R. Duthie, *Florists' Flowers and Societies* (Princes Risborough, 1988)'; 'Florist's or Select Flowers' in J. C. Loudon, *Encyclopaedia of Gardening, etc. A New Edition,* 5669–970 (London, 1834). See also www.auriculas.org.uk (2007).

9   'Manchester Floral Society', *Chronicle,* 14 April 1827.

10   Manchester Floral and Horticultural Society, *Manchester Times,* 8 August 1829.

11   Letter, 'A Florist', *Guardian,* 4 August 1827.

12   Advertisement, ' Manchester Floral Society', *Chronicle,* 14 April 1827.

13   Advertisement, ' Manchester Floral Society', *Chronicle,* 12 February 1831.

14   Correspondence, 'An Uninterested Observer', *Manchester Times,* 21 July 1832.

15   'Manchester Royal Botanical Gardens', *Manchester Faces and Places* 1 (Manchester, 1899), MBH 7/3/21.

16   See Letter to Stamford asking him to support the Manchester Society and its aims, 14 August 1827, EGR 4/2/10/20/2; also J. Lomax, 'The First and Second Earls of Wilton and Heaton House', *Transactions of the Lancashire and Cheshire Antiquarian Society* 82, 58–101 (Manchester, 1983).

17   A letter dated 19 May 1829 in the Enville archives makes it clear that subscriptions were not called for until the site for the Garden had been decided. See Letter, Hugo Worthington to the Earl of Stamford, G/2/2/8/9, GB-2184-Grey.

18   A. Prentice, *Historical Sketches and Personal Recollections of Manchester* (Manchester, 1851) (Chetham's Library, Manchester); T. Baker, *Memorials of a Dissenting Chapel* (Manchester, 1884); *Williams Deacon's 1771–1970* (Manchester, 1971).

19   See A. Brooks and B. Haworth, *Boomtown Manchester 1800–1850, The Portico Connection* (Manchester, 1993).

20   The Portico Library in Manchester is such an example, where women would only become members in 1882 after the passing of the *Married Women's Property Act.*

21   Letter, *Guardian,* 4 August 1827.

22   Bretton Hall is now the Yorkshire Sculpture Park.

23   Local Intelligence. *Manchester Times* 25 June 1831 and 26 May 1832.

CHAPTER FIVE

Creating the Garden

....

Liverpool's subscription botanic garden had opened in 1803; 27 years later
the Manchester Botanic Garden opened, aspiring to the same principles and
ambitions as its neighbour and rival though its larger site offered more delights
to the subscribers. June 1831 saw the Council well on their way to fulfilling
their original brief; satisfactory land had been acquired and a garden fulfilling
the design specifications of a botanic garden was under construction. Yet
already there were signs of forthcoming problems. The shortage of subscribing
members, at least in numbers that would provide the finance to run the Garden,
was already apparent. This 'incubus of debt' would persist and later Councils
would have to deal with the consequences.[1]

Before starting to explore the construction of the Garden, a brief examination
of the finances of the Society is important, as their approach to money and
their continuing debt was to prove a critical factor in the future development
of the Society. Reporting on 9 May 1829 on a 'thinly' attended meeting of the
Society, the *Courier* noted that the money in hand was £2679 6s 4d but, as
there were only 200 members, this would be insufficient capital to build the
proposed glasshouses. The Council had assured the *Courier* that they would
not begin construction until there was 'substantial evidence of a considerable
addition to their funds'. No complete membership list is available for this period
so the income of the Society cannot be estimated. In September 1831 Paxton
and Harrison published William Mowbray's design for the Manchester Botanic
Garden in the *Horticultural Register* and stated that 'the groundwork' had begun
two years previously, giving a date sometime in early 1829 as the start of the
building work.[2] Since they began in 1829, it must be assumed that either the

Section of the remaining rear boundary wall showing the brick piers of the back gate to the Garden, the end of Botanical Avenue.

Society had attracted many new members and could keep its promise given in May or, with high hopes for a successful venture, they were taking the first steps on the slippery slope of spending anticipated income. The finances will be a continuing theme throughout the following chapters on the Manchester Botanic Garden.

Reading the article it becomes clear that, since 1829, Mowbray's design had been executed on the ground, though it was still not yet complete in all its details. According to the *Courier* on 9 May 1829, the first decision taken by the Council was to secure the site by constructing the perimeter walls and to build the lodges, the latter a requirement under the purchase. The *Guardian* reported on 23 May that: 'The Council were prepared to treat for the execution of the BRICKWORK for a wall from 7–800 yds in length and for the supply of semi-circular and flagstone coping'. The contract went to Woodall, a local stonemason. Viewing the plan, it is apparent that the requirement for bricks was considerable, not only for the external walls but also for several structures within the body of the garden. One of the attractions of the land had been that the several clay pits to the south of the plot would allow bricks to be made on site at a saving to the Society. The tender for bricks reported on 9 May suggests that to allow work to start immediately the Council bought in initial supplies.[3]

A difficulty for the Council when supervising contracts was to be their chain of command. The Council had appointed as clerk of works Mr Shorland, a Manchester architect and surveyor, who would prepare plans and costings when required and would also liaise with Mowbray on site.[4] Neither man appears to have had the authority to deal with problems directly, the initiative always remaining with the Council, a reminder that both were employees of the Society operating within well-defined limits. Problems with the chain of command continued. To alleviate this, a Garden committee was finally established on 27 July and met regularly at the Garden to supervise the work. By 1830 plans

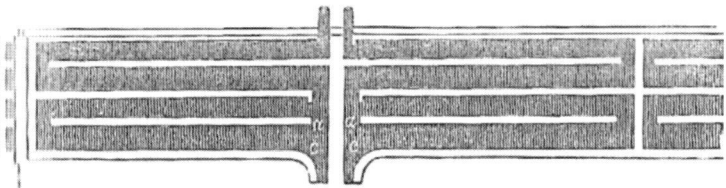

which may be thus described. In the wall (*fig.* 708.) there is an open space *c*, over the fireplace, where the smoke and heated air enter the flues; and here, at *a*, a damper is placed for regulating the admission of the heat throughout the whole wall. By drawing this damper a few inches, a certain portion, at the pleasure of the operator, of the hottest air, direct from the furnace, is allowed to ascend direct to the third flue, which thus renders the upper portion of the wall as hot as the lower part, without the necessity of any variation whatever in the construction of the brickwork. (*Gard. Mag.*, vol. viii. p. 670.) Flued walls may be heated by steam or hot water, as well as by smoke or hot air; but there is this great advantage of adopting steam or hot water, that regular flues are unnecessary, all that is required being to construct the wall hollow from bottom to top, introducing at the height of eighteen inches or two feet from the level of the ground a single tube for the circulation of the heated fluid.

Heating a flued wall.
From J. C. Loudon,
*An Encyclopaedia of
Gardening New Edition*,
ill. §3287 (London, 1834).

and estimates for the lodges had been sort and a contract awarded to Race and Bearsall, local builders. The Council sent them a letter on 16 June to the effect that, if work on the lodges was not finished by 24 June, they would enforce full penalties as stated in the contract.[5]

Plans were also well advanced for the construction of the glasshouses and Jones and Co of Birmingham had sent plans for alpine frames and the wings of the central range. In May the foundations were laid. Heating the range was an expense that the committee considered in mid-June. Smoke or the new hot water system – these were their chosen options. Jones and Mowbray were to supply estimates and Mowbray's plan for building the internal walls was also considered, as was the height and whether flued or solid, whether copings were needed and the amount of bricks required.

To give some estimate of the bricks used, in June 1829, 6000 bricks, 3 in by 9 in (76 × 229 mm) were made and fired in the grounds. During the period of construction the Committee had to purchase specialist bricks off site including bricks for the flues. Stone was obtained from local quarries and brought to the garden by canal; Catlow sandstone from Colne, flags from Greenwood's quarry at nearby Blackburn, and the rockwork stone from Greenfield, near Saddleworth. By September the back sheds and kitchens (which must have required considerable brickwork) were plastered and bricks for the conservatory, together with firebricks for the hot walls, ordered. Brick then disappeared from the agenda though work carried on until the beginning of December when the

Holly. From *The Woodland Companion*, pl. 28 (London, 1802). In 1831 the Society ordered 30 hollies, 'in sorts' from Caldwell's Nursery, Knutsford.

weather closed in. Construction continued after the winter, though the garden was not completely finished when it was opened to subscribers in June 1831: 100,000 common bricks were ordered from Mr Brownhill on 9 June 1831, at the price of 23s 6d per ton.

Building the hard landscaping and structures was not the only consideration. The very nature of the enterprise meant a considerable quantity of plants in large variety had also to be acquired. The main entries in the minutes referring to plants are acknowledgments of gifts and these donations to the Society continued throughout the Garden's existence.[6] The minutes over the years confirm that, as Loudon had predicted when describing his design for the Birmingham Botanic Garden, 'almost all the plants will be received in presents, or in exchange, from other public establishments'.[7] The earliest donation was a collection of willows from the Duke of Bedford on 8 March 1830.[8] In June the same year a member presented a collection of dahlias, which had become very fashionable after notable hybrids were bred in France in 1815.[9] This was possibly Edward Leeds, a passionate gardener, and long time subscriber to the Society. Leeds was famous for his varieties of Narcissus.[10] An advertisement in the *Manchester Guardian,* 8 September 1832, advertises Dahlias to order from his house in Stretford, where they could be viewed.

Birmingham, Liverpool, Glasgow, Dublin, and Edinburgh Botanic Gardens and Kew, all contributed. Seeds were sent from abroad, directly to the Society, or presented through friends in Britain, and these sources included New South Wales, Mexico and Brazil.[11] Donations were always acknowledged and the donor made an Honorary member. Only one instance has been found of the Manchester Society taking part in an expedition to find plants. In 1831, they acceded to William Hooker's request to part-sponsor Glasgow Botanic Garden's

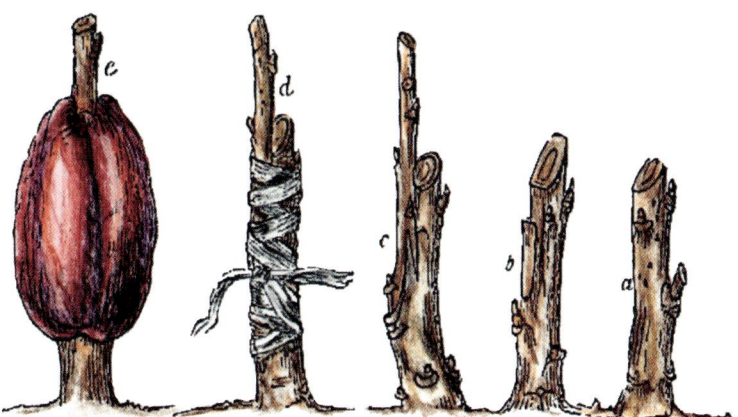

Splice-grafting in its different stages.

Grafting by detached scions. From J. C. Loudon, *The Horticulturist*, ed. W. Robinson, 250 (London, 1871).

expedition to Mexico.[12] Information was another asset that the Council was keen to receive. Mowbray was asked to make enquires of, or visits to, other gardens to see horticultural equipment in operation and, hopefully, to acquire more plants. The local press keenly followed these developments. In September the *Guardian* reported that the London Horticultural Society had offered the Committee cuttings of plants and 'a wealth of flower seeds'.[13]

There is no detailed planting plan for any part of the Garden but it is clear that a considerable number of trees were established at this stage as they were needed for the arboretum, the fruit garden, and as specimens in lawns and beds. On 21 December 1830, Mowbray presented the Council with a list of trees required for the arboretum and he was asked to write to Miller's Nursery in Bristol for an estimate.[14] Two members of the Council, Moore and Walker, submitted a fruit tree report on 29 December, which appended a list of those supplied by the London Horticultural Society for the Garden. An order to Caldwell's nursery, Knutsford in February 1831 shows how varied the planting was to be.[15] [This nursery supplied plants and seeds to many of the nobility in Cheshire and Lancashire, including Dunham Massey, Tatton Park, and Lyme Hall. Several Council members' names also appear in Caldwell's order books, including the Rev. John Clowes, William Bow, William Egerton (Tatton Park), and H. H. Birley.[16]]

Though a catalogue of plants is still available for the Liverpool Botanic Garden (and the minutes of 9 June 1830 show Mr Sergeant donated one to the Library at the Manchester Garden) no evidence has been found of a surviving list for the Manchester Garden.[17] Very occasionally lists of donations appear in the minutes. There are several in the years following the conversion of the Garden to a pleasure ground after 1848. This suggests that an appeal may have been made for donations of plants.

Botanical beds were included on the plan to encourage scientific studies. Plants were labelled in the Linnaean system and, as an aid to encouraging

Report of Messrs Moore and Walker 27 December 1830
Minutes, 27 December 1830, MBH 2/2/1

Agreeably to the request of the Council of the Manchester Botanical & Horticultural Society Mr. Moore and Mr. Walker submit the following report as the result of a conference they have lately had with the very able assistant Secretary to the London Horticultural Society regarding the experience of that institution in raising and distributing Fruit Trees. Fruit trees procured from Nursery men after having been carefully trained and pruned for several years, often turn out to be very inferior both in kind and quality to those for which they were purchased. It will be an important consideration with the Manchester Society to do away, as much as possible, with this vexatious loss of time and labour by furnishing its Subscribers with plants and scions upon which they may more safely depend. There is great reason to believe this desirable object may be soonest and most satisfactorily attained by attending to the following regulations.

1st.   Not to hasten the covering of the Walls nor ground by extensive purchases of Fruit Trees from nurserymen however strongly recommended

2nd.   Not to deliver under the sanction of the Society scions from any fruit trees before fairly mature fruit has been grown upon it in the Society's own garden.

3rd.   To keep the plants and scions furnished by *The London Horticultural Society* to the Society very distinctly marked and arranged.

4th.   To direct the curator to forward the engrafting and budding of the best sorts upon the stocks raised in the Society's own gardens and to pay the greatest attention to the registering of this work.

5th.   To request that with Noblemen and Gentlemen as are kind enough to offer plants or scions to the Society will be pleased to allow the curator or some of the Council to see the parent plants bearing fruit.

6th.   To take care that a constant and abundant supply of the most approved stocks for engrafting be kept up in the Society's garden.

John Moore C. J. S. Walker December 27th 1830

Memorandum: That the secretary of the London Soc. sent a list of fruit trees which had been sent to the Soc. and of cuttings which could be sent in Spring.

Fruit Trees Etc. forwarded:
40 Apples different sorts 20 Irises different sorts
13 Pears ditto 25 miscellaneous plants consisting rubus(3), Ribes (1), Spirea (7)
3 Plums ditto Prunus (2), Rose (5), Iris (7), 48 cuttings of Ribes (11), Pyrus (12),
4 Cherries ditto Craetagus (21), Mistfriclus (1), Sorbus (1), Cotoneaster (2)
50 Roses ditto
22 Paeonies ditto

Cutting which can be sent in Spring:
30 Apples diff. sorts 14 Peaches do. 19 Pears do. 5 Plums do. 5 Nectarines 4 Cherries do. 6 Apricots do.

Order to Caldwell's Nursery, Knutsford 26 February 1831.
DDX 363, Messrs Caldwell & Sons, Knutsford, Archives,
Chester Record Office, Chester

24 White Cedar 10/-
24 American Oaks in sorts 12/-
6 Gleditchia Tricanthos 2/-
14 Spirea 14 sorts 3/6
6 Deciduous Cypress 6/-
3 Striped Elm 1/6
2 Leucomb Oak 5/-
20 Sorbus hybrida 6/8
6 Pinus Glaber 3/-
12 Evergreen Oaks 6/-
6 Arbutus in pots 9/-
12 Alatenus sorts 8/-
24 Cytisus in 4 sorts 8/-
6 Candleberry Myrtle 3/-
12 Fernleaved Beech 18/-
3 Cutleaved Alder 1/-
24 Oakleaved Hornbeam 8/-
2 Yellow barked Lime 6/-
6 Robinias in 3 sorts 6/-
24 Maples in 6 sorts 8/-
12 Ashes in 4 sorts 6/-
2 Laurus Benzion 2/-
6 Pinus prazaria 1/6
6 Hemlock Spruce 6/-
12 Red Cedars Large 12/-
2 Tree Ivy 1/-
6 Myrtle leaved Box 2/-
24 Arbor Vitae 16/-
12 Balm of Gilead 3 feet 2/- (Poplar)
5 Double Thorns 5/-
2 New Scarlet do 3/-
1 tru (sic) Shipleys Apricot 4/0
2 Sea Buckthorns 2/6
6 White Mulberry 3/-
100 Large Hornbeam 12/6
Carriage 15/-

12 Swedish Juniper 4/-
18 Craetagus 3 sorts 9/-
30 Bird Cherry 3/6
6 Black Birch 1/6
6 Exeter Elm 3/-
24 Acacia
50 Sorbus Domestica 7/6
2 Butchers Broom 1/-
6 Pinus Cembra 9/-
3 Quercus bullata 3/-
12 Laurenstinus 6/- sorts
10 Shrubby Honeysuckle 5 sorts 10/6
6 Red Gale 2/-
6 Prunus chicasa 3/-
3 Striped Beech 4/6
2 Rheumus Latifolius 1/-
12 Virginia do. 4/-
3 Paper birch 1/-
2 Purstam? tree 1/-
25 Std Snowy Mespilus £1/1/8 (Amelanchier)
6 Yellow Jasmine 1/6
3 ptelea Trifoliata 1/6
6 Euonymus in 3 sorts 1/6
6 Savuis? Large 4/6
30 Hollies in sorts 20/-
6 phyllyreas in sorts 4/6
1 Yucca Augustifolia 3/6
50 Spruce Fir 3 feet 10/-
25 Large Birch 6/3
5 Old Scarlet do 5/-
12 Silver Fir 2/-
3 Dwf Siberian Elms
2 prinos? Virticillata 1/-
6 Halesia tetraplara 3/-

Total £19 3s. 1d.

This is a total of 622 trees, deciduous and evergreen, together with shrubs and fruit trees.

Plant Accessions

The following lists give an impression of the scope of the plants coming into the Society from both home and abroad. [*Spelling as original though arranged alphabetically*]

2 October 1851, Mr. Moore, Botanic Garden, Chelsea

Aloes: 16 species, Echeverias 3 species, Elphast tree 3 cuttings, Euthalis Picta, Gustinius 6 species, Hawthorns 11 species, Hoya Cunninghamii, Isoloma mollis, Other cuttings.

2 October 1851, Mr. Marnock Botanic Garden Regent's Park

Achinea Fulgens, Cantica bi-colour, *Cantura pyricifolia, Crinum seabrum, Hedysanium gyzans, Hymenea courbunil, Pixtia shatiotis, Pontedaria cassifolia, Pontederia punica, Tamarindus indica, Zillandaia zebrine purpurea,* Some tropical and hardy ferns. 260 species herbaceous plants and grasses, Seeds *Victoria Regia.*

5 February 1852, J. Rauch, Wien, Austria through Mr. Bellott, Surgeon, Manchester

Acer obtusatum, Allium Suareolius, Allysum montanum, Androsacae elongata, Androsace obtusifolia, Aquilegia erisioxa, Arenasia Austriacus, Arenaria Rhapontica, Aria Schostii, Callianthemum anemonoides, Callianthemum intrafolium, Campanula barbata, Campanula aplina, Centurea Phrygia, Campanula Sibirisa, Cytisus Weldenii, Dianthus alpinus, Gentiana puctata, Heiracium alpestre, Hieracium rupestra, Linum Au[n]driacum, Phytenuma paucifolia, Primula longiflora, Primua villosa, Scabiosa holosencia, Sempervivum tintorum purpureum, Serratula radiata, Sivertia punctata, Soldanella Montana, Verbascum floressium, Viola Montana, Wullfinia lariniliaca.

MBH REPRODUCED BY COURTESY OF THE UNIVERSITY LIBRARIAN AND DIRECTOR, THE JOHN RYLANDS UNIVERSITY LIBRARY, THE UNIVERSITY OF MANCHESTER

The Tamarind. From *Vegetable Substances Timber Trees: Fruits* (London, 1829).

Soldinella Montana. From J. Wood, *Hardy Perennials* (London, 1884).

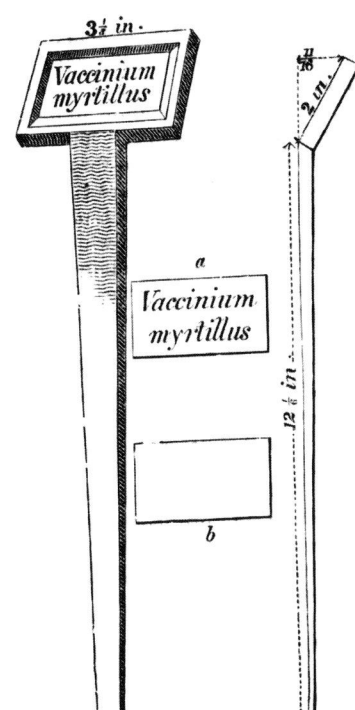

FRUITS.

The double daggers (‡‡) designate such fruits as are new, and have not been sold previously to the printing of this Catalogue.

APPLES.

*Those marked * are planted principally for Cider.*

1 Ashmead's Kernel
2 Beauty of Wilts
3 Best Pool
4 Borstorff
5 Brandy
Golden Harvey
6 Bur-stock
7 *Cadbury
8 Calville, white
9 Cam House
10 Cat's-head
11 *Coccagee
12 Codlin, Carlisle
13 ——, Dutch
Glory of the West
14 —— Early Downton
15 ——, English
16 ——, Keswick
17 ——, Manks
18 ——, Nelson's
19 ——, Transparent
20 Cornish Gilliflower
21 Court of Wick
22 Court-pendu plat
Garnons.
23 De la Ware
24 *Devonshire Wilding
25 Doxey
26 Dredge's Emperour
27 Duke of Wellington
28 Dutch Mignonne
29 Duchess of Oldenberg
30 Early Leicester
31 —— marrow
32 —— New-York
33 Emperor Alexander
34 Fenouillet gris
35 flower of Kent
36 *Fox-whelp, red
37 golden jennet
38 —— Leadington
39 Gravenstein

APPLES.

40 Greening, Rhode Island
41 ——, Yorkshire
Lord Chency's
42 Hagloe Crab
43 Harvey's
Doctor Harvey's
44 Hawthornden
45 Hedging, summer
46 Herefordshire Goose-apple
47 *—— underleaf
48 Hicks's fancy
49 Jaunette
50 Jersey
51 Ingestrie, red
52 —— yellow
53 Juneating, red
54 ——, white
55 King of the Pippins
56 *Knight's Grange
57 Lady de Grey
58 L'oignonette
59 Lucombe's Pine
60 Margaret
61 Margill
62 Mary Greed's
63 Moggs's long-keeper
64 *Morchin's crab
65 *Moreton
66 *Mortimer
67 Nonpareille
68 ——, American
69 ——, Braddick's
70 ——, early
71 ——, Martin
72 ——, scarlet
73 Nonsuch
74 Norfolk Beanfin
75 —— Paradise
76 Old English, *sent to America in 1636*
77 Ord's
78 Ortley
79 *Parsons's Ambrosia
80 *—— long-hanger

Left: Murray's named tally. From J. C. Loudon, *An Encyclopaedia of Gardening New Edition*, ill. §2208 (London, 1834). AUTHOR'S COLLECTION

Right: Page from the catalogue of fruits available from Miller's Nursery 1830 (reprint). AUTHOR'S COLLECTION

botanical studies, Loudon had recommended that plant names should be painted in large letters on the bevelled face of a cast iron tally so as to be legible without stooping.[18] The tally would give the name, class, native country, and time of flowering. The minutes record that Mowbray was asked to order two sample tallies from a joiner on 4 August 1830 and that 1500 were ordered four days later, suggesting they were wood rather than metal.[19] This quantity must have meant that not only the botanical beds, but also the plants throughout the garden were labelled for the instruction of the subscribers.

The opening of the Garden was not mentioned in the minutes and no description has been found in the local press; a reference in the minutes of 22 September 1830 showed a committee had been established to plan the

The Orange. From
*Vegetable Substances
Timber Trees: Fruits*
(London, 1829).

opening. The *Chronicle* noted on 4 May 1831 that during the Winter and Spring the Council had adopted means to add to the 'beauty and variety of such an agreeable retreat' and that the grounds already presented a delightful scene. Another entry in the minutes at the beginning of June 1831 indicated that they were anticipating the coming event. The Council arranged for twelve placards to be ordered and placed in different parts of the garden: 'Visitors are particularly requested not to walk on this grass or border'; possibly the first record of this ubiquitous notice?[20] A notice in the *Guardian* on 14 May, under the heading 'Manchester Botanical and Horticultural Society' also reported that public admission to the Garden would cease after 24 June.[21] The Society held its first exhibition of the year in the Exchange Dining Room, not the Garden, on 20 June 1831. Plants from the Garden were placed in a separate display and included an orange tree and a *cactus speciocissima* in full bloom.[22] The *Guardian,* on 18 June, had reported that exhibitors wishing to send material to the London Horticultural Society's Annual Exhibition on 22 June 'would receive every assistance in forwarding them', perhaps indicating that the Garden was not open by 20 June. Given all this information, it would seem likely that the Garden was officially opened on 24 June 1831. The members who attended the opening would have had a major disappointment. Only the outer wings of the glasshouse, a major attraction for exhibiting exotic plants, were finished. The central conservatory was missing. Had the *Courier's* report of 1829 come true?[23] Was there insufficient capital? How many members had they attracted?

Notes

1 Quoted in William Robinson, 'Notes on Gardens – No. XVII Manchester Botanic Garden' (*Gardener's Chronicle,* 24 September 1864, 915–6).

2 *Horticultural Register,* 1 September 1831, 107. Minutes for the Society do not start until 1830.

3 The main supplier of bricks was Mr William Holland, bricksetter, 20 Camp Street, Manchester: *Pigot's Manchester Directory* (Manchester, 1830).

4 George Shorland had an office in the Town Hall and lived at 16 Lower Mosley Street: *Pigot's Manchester Directory* (Manchester, 1832).

5 The only detailed example of full contract terms extant in the minutes occurs on 19 August 1830. Mr Crowther was asked to apply coping to the fruit garden wall in accordance with his estimate of 3s 9d a yard. The contract stated: 'Agreed 100 yds in three weeks the rest to be finished in five weeks or a £20 penalty, two thirds to be paid on the arrival of the stone. Any coping at 2ft 4 inches as opposed to 19 inches on the same terms, any 12 inches at 2s 2d per yard. Circular coping for the yards of the Lodges 18 inches by 3 inches at 3s per yard'. Crowther had to supply the cement and was supervised by Shorland.

6 Botanical books were also donated to the Library. Examples included: W. Forsyth, *A Treatise on the Culture and Management of Fruit Trees, etc., Vol. 1* (London, 1st edition published 1802); *Catalogue of Plants in the Botanic Garden at Liverpool* (Liverpool, 1808), and J. Hill, *The Construction of Timber from its Early Growth, Explained by the Microscope and Proved from Experiments, etc.* (London, 1770).

7 J. C. Loudon, 'The Birmingham Botanic Garden', *Gardener's Magazine,* 8, 427 (1832).

8 Plants and seeds were also sent from: Trentham (Duke of Sutherland), Tatton Park (Lord Egerton), Blithfield (Lord Bagot), Woburn Abbey (Duke of Bedford), Bretton Hall (Mrs Beaumont), Wentworth House (Earl Fitzwilliam), Croome Park (Earl of Coventry) and Welbeck Abbey (Duke of Portland).

9 D. Stuart, *The Garden Triumphant: a Victorian legacy,* 162–3 (London, 1988).

10 *Gardener's Chronicle,* 10 November 1894.

11 The Portico Library's visitor's book shows many merchants came from overseas at the same time each year including from North and South America; A. Brooks and B. Howarth, *Boomtown Manchester 1800–1850: The Portico Connection,* 109 (Manchester, 1993).

12 Minutes, 16 February 1831, MBH 2/1/1. They received the seeds and Minutes indicate correspondence with Hooker, anxious to find if they successfully germinated.

13 *Guardian,* 27 September 1828. Repeated by the *Courier* on 29 November 1828.

14 John Miller & Co owned a nurseries at Durdham Down and Arnos Vale, Bristol: *Pigot's Directory* (Bristol, 1830). Richard Forest, of Miller & Co, designed Bristol Zoological Gardens in 1835, a subscription garden for the study of natural history, that was also used for social events and flower shows

15 Messrs Caldwell & Sons, DDX 363, Cheshire County Archives, Chester Record Office, Chester.

16 No orders have been found in the Minutes for Manchester nurseries during 1829–1831 though several were listed in the local directories.

17 The Minutes of 22 November 1832 show that the apprentices were instructed to draw up a list of plants in the garden to send to Mowbray's successor, Alexander Campbell.

18 J. C. Loudon, *Encyclopaedia of Gardening,* §2043 (London, 1822).

19   See also Article XV, 'Description of a new Tally for naming plants', Letter, Stuart
      Murray, Curator Glasgow Botanic Garden, *Gardener's Magazine,* September 1827,
      28–9, with illustration; (Robert Marnock), 'Botanical Labels', *Florist's Journal,* 1840–1,
      159–63; F. Hanham, *A Manual for the Park,* ix (London, 1857). This gives a detailed
      description of a tally in the Bath Botanic Garden.
20   Minutes, 2 June 1831, MBH 2/1/1.
21   *Guardian,* 14 May 1831. Again the use of Public means non-members.
22   *Chronicle,* 25 June 1831.
23   *Courier,* 9 May 1829. The Society's governing body was variously referred to as
      'Committee' or 'Council' and the latter term will be used throughout the following
      chapters except in direct quotations.

CHAPTER SIX

The Glasshouses
'Who loves a garden loves a greenhouse too.'[1]

Lean-to cool house.
From J. C. Loudon, *The Horticulturist*, fig. 154 (London, 1842).

The *Chronicle*'s reporter attended the annual general meeting in January 1830 and noted: 'the Institution … is proceeding most successfully in its career' and further noted that the membership was not restricted to 400 as was popularly believed, but that 'unlimited numbers' could join.[2] Subscribers at this date were paying 1 guinea. The *Chronicle* warned that to participate from June 1830 the privileges were to cost 2 guineas and considered that: 'on these liberal terms the subscription list will be speedily increased'. Potential members had also been warned of a further hurdle. A year earlier the *Guardian* had informed readers that from 2 June 1829 new members were to be balloted and those who subscribed earlier 'would not have to undergo this ordeal'; in fact the balloting balls were not ordered until 22 February 1830. With the building costs now escalating, attracting new members was crucial. The Council seems to have anticipated an increase as they embarked on the costly glasshouse projects recorded in 1830 and 1831. Unfortunately there are no records of membership at this date. However it would seem safe to assume that the Committee decided that without the glasshouses the garden would lack its central attraction. They therefore proceeded in the belief that glasshouses full of exotic plants would draw in sufficient new members to cover the cost. In March 1830 when the

Mrs Beaumont's
Conservatory, Bretton
Hall, Yorkshire
(*Gardener's Magazine* 5
(1829), 681).
© ROYAL BOTANIC GARDEN,
EDINBURGH

tendering process was underway, the Council set up a separate account for the glasshouses, planning to use half of all future receipts until the funds reached £4000. This seems to confirm that the houses were built in anticipation of the money becoming available – the first step on a slippery financial slope.

Greenhouses and their designs, together with possible suppliers, were important discussion topics for the Council as these would be the most expensive structures to be built in the Garden.[3] Continued advances in both iron and glass technologies had widened the choices open to the Council and tenders were invited for the supply of the glasshouses. When the minutes of the Society start on 25 January 1830 the Council was considering several designs for forcing houses, conservatories, and hothouses.

A visit to Mrs Beaumont's famous conservatory at Bretton Hall in Yorkshire was proposed for the Curator, William Mowbray, and the clerk of works, Mr Shorland, to see how it was constructed and heated. (In 1832 on her death the dome was offered to the Society for 750 guineas but proved too expensive to move. It was finally sold at auction).[4]

On 1 February 1830, four sets of drawings, each with a plan and view of the elevation, were examined and rejected. Tenders were sent by Cottam and Hallam (Manchester), Jones and Co (Birmingham), Mr Week, and Mr Clark (addresses unknown) and each tender apparently contained several designs though it is obvious from the minutes that Jones was the preferred contractor from an early stage. Jones had already built glasshouses in the region for members of the Society. The following week the Council wrote to Jones asking for the prices of various methods of construction and heating, stressing the importance of the cost of the three varieties of glass to be used: upright, flat-roofed, and curvilinear. The new design was still too expensive so the Council asked for further plans.[5] A new contender emerged when the Council contacted

Bailey's of Great Bridge, near Birmingham. They had a national reputation as glasshouse manufacturers to the aristocracy.[6] On 8 February 1830 the Council were still asking for new tenders, quoting costs of the different forms and quality of the glass. At the Council's suggestion Shorland drew new plans claiming not to have used those previously submitted; in fact they used Jones No.1 Design as the model.

The Council's indecision led to complaints and disputes. After being sent another invitation to tender, William Bailey vehemently protested that he had been assured of a conclusion 'in his favour'.[7] Bailey's letter of protest to the Council made charges that 'deeply involved the honour of their characters'; Bailey later sent an apology to no avail. Still in doubt, the Council contacted Mr Cooper, Head Gardener at Wentworth House, Sheffield, on 10 March 1830, asking his opinion on the difference curvilinear and flat roofs made to the culture of plants. Finally the Council came to several decisions. A curvilinear roof was to be used only on the conservatory due to the high cost involved, and building would commence on one wing only. The special fund was then set up for building the glasshouses.

On 23 March 1830 Jones' tender was accepted and the contract, which was drawn up on the 26 April, also included a clause stating that Jones would continue to keep the houses in good repair. The work, which did not include the central conservatory, was to start on 1 June and be finished, to Shorland and Mowbray's satisfaction, by 1 August. Much to the Council's concern Jones proved dilatory, finally arriving on 5 July. The foundations for the stove, the connecting corridor, and the greenhouse were laid out in the next five days.[8] As the stone for the next foundations had not arrived, Jones planned to start them on 16 August. Over the months the Council's discussions on heating the houses had been protracted and they finally decided, after some opposition, to adopt the new hot water method, rather than use fires and smoke, even though it was more expensive.[9]

At the beginning of November the boiler had not arrived and the Council threatened to charge Jones for the loss of any plants due to frost if it was not installed at once. The boiler must have been installed as Jones finally arrived at the Garden on 23 December 1830 to explain how the hot water system worked. In April 1831 the Council sent Jones further modifications to their plan for the glasshouses, including the addition of another stove and wing, asking for an estimate of the possible costs. Jones attended a meeting on 28 April and quoted a price of £850–£1000 for the new work inclusive of hot water or, if they preferred, he would supply the boiler, and the pipes could be cast in Manchester under his supervision. On 5 May, with the opening of the Garden imminent, there was a failed attempt to arrange a special meeting with Jones. The Council threatened that others were interested in the work though the minutes do not name them; perhaps this was a tactic on the part of the Council to influence Jones. Finally, on 30 May, Jones agreed he could begin the projected new work in June and complete by the end of September.

It was during this period, in July 1831, that the acknowledged gardening guru of the day, John Claudius Loudon, visited the Manchester Garden and complained that it had 'a sameness' which did not please him.[10] According to Loudon, the planting of the rockwork was less than satisfactory as showy plants were distributed indiscriminately, making the scheme monotonous rather than, as he preferred, placing 'every genus by itself in an irregular group' for impact. Loudon, who does not mention the incomplete glasshouses, reserved his major complaint for the Arboretum where he objected to the nurse trees being a common mixture instead of the same genus as those they protected (Manchester was in distinguished company as Loudon had criticised tree planting at Alton Towers, Trentham, Chatsworth, and Heaton Park). Mowbray was outraged, immediately sending a letter in reply. He began on the offensive: 'Although I never expected you would write any flattering account of our garden, I did expect a fair statement of what was done, an outline of the divisions of the garden, and the progress made, &c'.[11] Why would Mowbray expect a less than flattering account? A letter to 'Domestic Notices' in the April edition of Loudon's *Gardener's Magazine,* apparently showed the impact Loudon's views were having on Manchester's readership: '*The Garden of the Horticultural and Botanical Society of Manchester* is proceeding apace. … Two lodges are built, which, I am afraid, you will not like; and unfortunately, the main entrance to the garden is from the north (on the Chester road), which you have shown in your *Encyclopaedia of Gardening* to be always bad. – *Y. H. March 2, 1830*'.[12]

The Council and the Curator must have awaited Loudon's proposed visit with some trepidation when, even before it was finished, a supposedly local gardener was pointing out the Garden's design faults through the medium of such a widely-read magazine. Mowbray, not to be cowed, concluded that '… after all the pains which we have been at, to be thus misrepresented is, I think, a very hard case'.[13] It can be no surprise that Mowbray's article describing the Manchester Botanic Garden appeared in a rival publication, *The Horticultural Register*.[14] Joseph Paxton and Thomas Harrison, the editors, contradicted Loudon by reporting that 'The situation Mr. Mowbray has chosen for the kitchen gardens, as well as the disposition of the arboretum, water &c., do him the greatest credit'. They specifically lauded the rockwork, and concluded: 'Indeed we have no hesitation in saying, that it is by far the best laid out garden at present extant'. The Society's exhibition on 5 October 1831 at the garden, attracted '900 individuals of the highest respectability', a sign that the Council's targeted subscribers were supporting the Botanic Garden.[15]

A *Guardian* report on 10 September 1832 gives us a contemporary impression of the newly opened Garden:

> **Opening of the New Stretford Road** … A train of about sixteen carriages, gigs and horses, and headed by a band of music … reached the garden about twelve o'clock … The arboretum had grown on last year and in one or two years should provide shade. There were American Plants in bog soil. Of most notice was the walk over the rustic bridge through the rockery which every year will appear more interesting

to the lovers of wild scenery [with] mosses on the sides of huge stones, and heath and mountain shrubs rearing their heads. The borders were over but a monthly rose was flowering on the wall and the dahlias were dazzling.[16]

From the beginning the use of the garden as an amenity for the members as well as a place for botanical study had been apparent and although there was still work to be done the Society seemed set for a successful career.

Still the main concern for the Council must have been to finish the glasshouses to keep membership interest alive and attract further subscribers. In June 1832 the Council had invited estimates for completing the stonework on the left wing greenhouse, for the stove and greenhouse of the right-hand wing, and for the heating apparatus and its installation for the completed range.[17] When the Curator, Willam Mowbray, died in July 1832, the Council had other concerns; foremost was finding his successor. Alexander Campbell was appointed on 8 November 1832; his previous employer was an Honorary member, the Earl of Mount-Norris of Arley Hall, Bewdley. When he arrived his first task was to supervise the construction of the conservatory, the central glasshouse of the partly-finished range of hothouses. This was to be built to the design specifications, which had appeared in the minutes of 15 February 1830:

> That the dome of the Conservatory be 35ft high and not contain in the whole a greater number of superficial feet of glass than are shown in Jones and Co. plans No.1. That the final two wings to be 60ft. each in two compartments with a glazed corridor between both the Conservatory and the stoves and that each of the latter to be 40ft long in one compartment only – and that the whole be on the curvilinear principle and construction of iron exclusive with the exception of the doors which in Mr. Shorland's opinion would be best if in wood.

Conservatory and Houses, Erected 1832. Frontispiece to the *Catalogue of Plants cultivated in the Birmingham Botanic Garden* (Birmingham, 1836).

The archives of Clark and Jones (later incorporated into Chance Glassmakers) have an order for the central house of the range, the Conservatory, in their order book on March 1835.[18] Unfortunately the description of the order stops at 'back of roof', nor was a drawing included unlike most other specifications in the order book. Jones's estimate for the costs, £2350, was on 5 March 1835.[19]

This drawing of Jones's glasshouse for the Birmingham Garden was constructed on similar lines, though curvilinear throughout, and the illustration may give some impression of the appearance of the Manchester Conservatory.

Clark and Jones agreed to build a structure that would last 7 years at least and would include a ventilation system and all fittings except the heating, brick, and stonework. On 12 March work proceeded and the minutes include the price for the glass required: £5. Though no illustration of the conservatories at the Botanic Garden have been found, a description of the completed range was given in Benjamin Love's guide book, *Manchester as it is*; one of the few extant descriptions of the Botanic Garden.

> [The] plant houses and conservatory form a beautiful and imposing range of glass, three hundred and twenty-one feet in length, constructed on the curvilinear plan by Messrs. Clark and Jones of Birmingham, and heated by hot water in large pipes. The conservatory is in the centre of the range, about forty feet high, with a spacious dome-shaped roof, terminating in a point.[20]

Heating by hot water was a new technology and there had been much argument among the Council members about adopting the system on the grounds of costs. The drawbacks of the alternative, heating by smoke, were that the boiler had to be stoked constantly for 24 hours a day and the apprentices and a supervisor would need to be in attendance to keep the fire going.

By 1836 the Council could congratulate themselves on their success, the Conservatory, the foremost attraction to subscribers, was built and so they could look forward to the future with confidence.

Plan of a hot water system, Jones and Co. No drawings have been found of the Manchester system but a plan for a hot water heating system supplied to a Mrs Fisher by Jones and Co in 1835 demonstrates the principles involved. The boiler in this case was underneath the floor. The pipes had floor gratings covering them to allow the heated air to rise above.

HEATING SYSTEM FOR
CONSERVATORY, 13 JANUARY 1835
(FOR MRS. FISHER). 1056/249, CHASE
GLASSMAKERS ARCHIVES, ARCHIVES,
BIRMINGHAM CITY LIBRARY

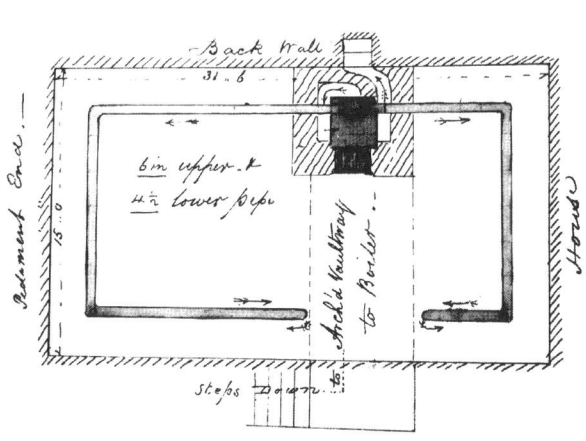

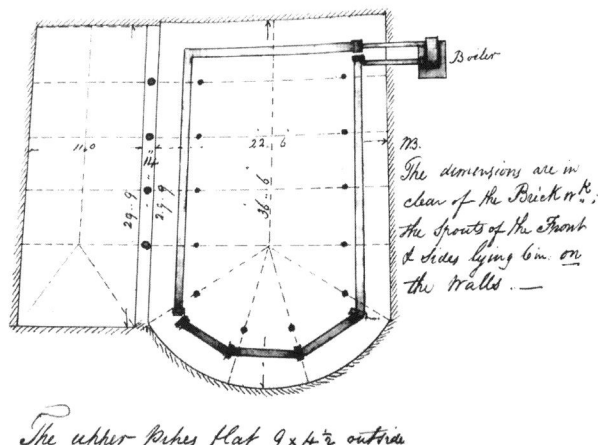

JAMES SENDALL & Co., Ltd., CAMBRIDGE.

CONSERVATORIES PLANT-HOUSES, ORCHID-HOUSES, VINERIES,

PEACH - HOUSES,

FORCING-HOUSE, &c,
Erected in any part of the Kingdom.

Surveys made, and Plans prepared to suit any position or requirements.

Catalogues and Estimates Free.

THE CHAMPION CHECK-END SADDLE BOILER.

Durable, Economical, Easily Fired, and Lowest in Price of any Boiler of its power.

ᒋist and Testimonials Post Free.

No. 100.

MELON OR CUCUMBER FRAMES.

2 Light Frame, 6 ft. by 4 ft.	..	..	..	£2	0	0
2 ,, ,,	8 ft. by 5 ft.	*Notice the*	2	12	6
2 ,, ,,	8 ft. by 6 ft.	*useful sizes*	2	15	0
3 ,, ,,	12 ft. by 6 ft.	*we make.*	3	15	0
4 ,, ,,	16 ft. by 6 ft.	..	..	4	15	0

AMATEUR'S FORCING HOUSE.
Tenant's Fixture

GARDEN FRAMES made and Stocked in all the most useful sizes. They are made of well-seasoned red-wood Deal, malleable iron-hinges, &c. PAINTED Three Coats of best Oil Colour. GLAZED with 21-oz. English Glass. NEW LIST POST FREE.

All Frames Packed Free, and Carriage Paid.

Cheapest and most useful House made at the Price.

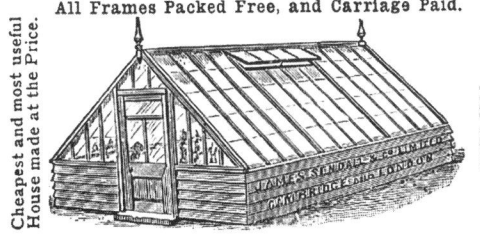

NEW ILLUSTRATED LISTS, POST FREE.

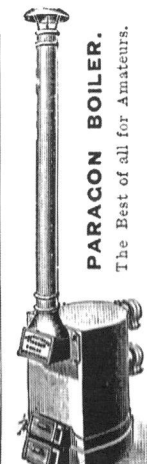

PARACON BOILER.
The Best of all for Amateurs.

This is an **EXACT VIEW** of the No. 78—**SPAN-ROOF FORCING HOUSES**, 2 feet 9 inches high at the sides, Ventilators on both sides of the roof, door at one end, with lock and brass hinges, made of the best red deal, carefully fitted so that anyone can put it together **IN ONE HOUR.** 21-oz. **ENGLISH MADE** Glass cut to size. Painted twice.

PACKED FREE AND CARRIAGE PAID.

Length	Width	Ridge		Price
15 feet	8 feet	7 feet	.. .. ..	£9 10 0
15 feet	10 feet	7 feet 6 in...	..	13 0 0
15 feet	12 feet	8 feet	.. .. ..	15 0 0

Advertisement showing a selection of glasshouses available. From J. Birkenhead, *Ferns and Fern Culture* (Manchester, n.d.); Author's Collection. Birkenhead had a fern nursery in Timperley, near Manchester.

Notes

1 W. Cowper, *The Task, A Poem in Six Books, Book III. The Garden* (Albany, 1810).

2 *Chronicle,* 30 January 1830.

3 J. Loudon, *Encyclopaedia,* §§ 2450–540 (1834). This covers glasshouses in great detail.

4 *Minutes,* 18 June 1832, MBH 2/1/1. The conservatory is listed in the catalogue as comprising lots 61–103 inclusive and lots 122–39 inclusive. A copy is in the RHS Lindley Library, London.

5 Though no examples of the Manchester tenders have been found, Jones's quotation for the Greenhouse, Stove and Conservatory for the Birmingham Botanic Garden was £1650. Estimate, John Jones and Co, Mount Street, Birmingham, M/S 1520/55/5, Archives, Birmingham Botanical and Horticultural Society, Birmingham City Library.

6 Bailey's had not been mentioned in the original list of tenders. Bailey had erected glasshouses at several aristocratic houses in Britain including Syon House and Alnwick for the Duke of Northumberland. He had also built a spectacular curvilinear greenhouse (extant), at Bicton Park, Budleigh Salterton, in the early 1820s.

7 Letter, Minutes, 26 February 1830, MBH 2/1/1.

8 Throughout the garden's construction many delays had been caused by the reporting system instigated between the Council, Shorland and Mowbray and at this point the Council authorised Mowbray to supervise work if Shorland was absent.

9 See J. Loudon, *Encyclopaedia,* §§ 2527–40 (1834).

10 *Gardener's Magazine,* August 1831, 143.

11 'Retrospective Criticism', *Gardener's Magazine,* September 1831, 616.

12 *Gardener's Magazine,* April 1830, 334.

13 *Gardener's Magazine*, September 1831, 557.

14 *Horticultural Register*, 1 September 1831, 109.

15 *Chronicle,* 27 September 1831.

16 *Guardian,* 10 September 1832.

17 *Minutes,* 21 and 28 June 1832, MBH 2/1/1.

18 O/N 1132 1835, Manchester Botanical and Horticultural Society, Chance Glassmakers, formerly Jones and Clark, then Jones, 1056/249, Archives, Birmingham Library, Birmingham.

19 *Minutes,* 5 March 1835, MBH 2/1/2.

20 B. Love, *Manchester as it is*, 121–2 (Manchester, 1839). According to the archives of Chance glassmakers, Thomas Clark founded the firm of Clark in premises in Lionel Street, Manchester in 1818. Could it be the two men met when Jones first came to work in Manchester and later combined to found Clark and Jones?

CHAPTER SEVEN

Staffing the Garden

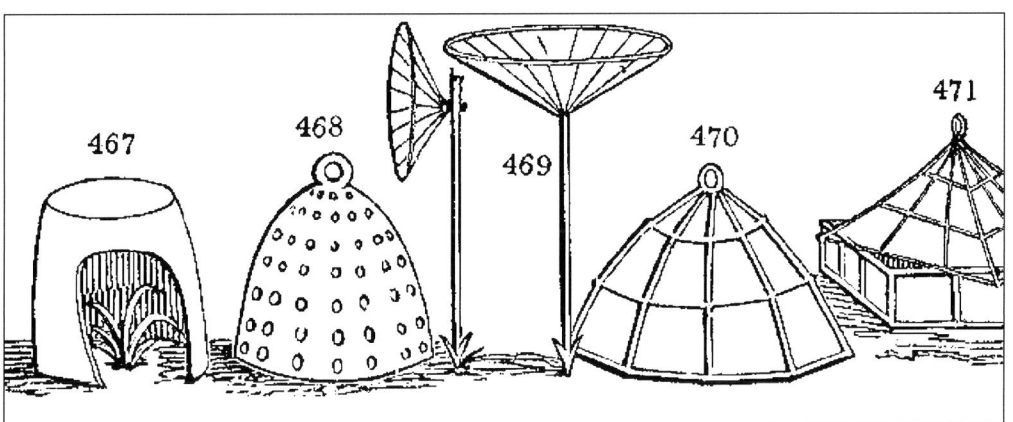

467   468   469   470   471

Hand-glasses are
of various species.
From J. C. Loudon,
*An Encyclopaedia of
Gardening New Edition*
§2266 (London, 1834).
AUTHOR'S COLLECTION

When the Council selected Alexander Campbell in 1832 they were appointing the most important member of staff of any subscription botanic garden as the character of the curator played a major part in the success of each society. These early curators had originally worked for wealthy individuals, many of whom had their own private botanic gardens. Later curators had often been trained at Kew and this, together with previous experience at a subscription botanic garden, was to appear on their *curricula vitae*. Some curators became nationally known figures in the gardening world. For example Robert Marnock, Curator at the Sheffield Botanical Garden (1834), became Curator of the Royal Botanic Garden's gardens in Regent Park (1841–1869). He also became a well-known garden writer and landscape gardener. Each curator lived at their Garden, supervised the maintenance and the staff, and took their instructions from the Council of their Society. Loudon lists the duties of the curator of a public botanic gardens as: creating a herbarium, collecting wild plants, acclimatising plants, distributing seeds, cuttings and plants, knowing the names and histories of the flora in the garden, and disseminating and distributing rare plant specimens.[1]

Examination of Campbell's appointment shows that on 25 October 1832 the Council were examining letters from the applicants; no names were given and no advertisement has been found. The terms of the appointment were £100 per year plus house, rent, and property tax paid, with the provision of coals and candles; a rise of £20 on Mowbray's wages. On 8 November Mr Alexander

An Aside: Alexander Campbell, Curator 1832–1857

Born in 1785 at Contin, Rossshire, Alexander Campbell began his gardening apprenticeship at Braham Castle, Nairn, home of the Earls of Seaforth.[1] In March 1815 he moved to Bookham Grove, Surrey, a private botanic garden with tropical plants. A year later he moved to Claremont, Surrey, the home of Princess Charlotte and her husband Leopold, Prince of Saxe-Coburg (later King of the Belgians) to take charge of the plant and forcing departments. Princess Charlotte's untimely death in 1817 saw Campbell move to the Royal Botanic Garden, Kew. 1818 saw another move, this time to serve Mrs Beaumont at Bretton Hall in Yorkshire. In 1819 he move to the Horticultural Society of London, where he was appointed foreman of the gardens. In 1821 he moved to the garden of the Comtesse de Vandes, Bayswater, where Loudon claimed in 1826 that 'more new plants have been published in botanical works as having first flowered here, than in any private garden around London'. Like Mowbray, Campbell's experience was mainly in private gardens. Again like Mowbray, his skill had been acknowledged by the gardening world. In 1828 he was awarded the Banksian Medal by the Horticultural Society of London, and a second recommendation noted 'his skill in the cultivation of Stove plants and particularly of *Combretum comosum**, which had not flowered in any other collection'.[2] Campbell was a member of the Horticultural Society of London, an Associate of the Linnaean Society and the Medico-Botanical Society of London. When the Comtesse de Vandes died in 1832, Loudon commented that ' We trust some good is waiting Mr. Campbell … there is not a more amiable and worthy man, or better gardener'.[3] Later that year he was appointed Curator of the Manchester Botanic Garden.

After working for the Society for 25 years, Campbell retired as Curator in1857 and from the Society's employ in 1864. In his obituary the correspondent of the *Gardener's Chronicle* compared him to a living book of reference, a passionate plantsman who welcomed friends and strangers alike to the garden and that many new and rare plants were grown to enrich the Garden he loved:

How many grown-up amateurs have dated their first knowledge of plants from Mr. Campbell may never be known, but the seeds were sown, and we are happy to say the crop is heavy, for the gardens within thirty miles of Manchester are to this day rich in floral wealth.

Alexander Campbell was aged 82 when he died a peaceful natural death at his home, 11 Thyrsa Street, Chorlton Road, Manchester in 1877.

Combretum comosum: Evergreen climbing stove plant. Introduced from Mexico, Trinidad, Brazil and the West coast of Africa from 1818 onwards. From G. Nicholson, *The Illustrated Dictionary of Gardening, Div. II* (London, n.d.; Author's Collection

[1] A. Forsyth, 'Obituary (for Alexander Campbell)', *Gardener's Chronicle,* 27 January 1877, 119. All quotations are taken from this source unless specified.

[2] *Transactions of the Horticultural Society of London* 7 (1830).

[3] 'The Botanical Collection of the late Comtesse de Vandes, at Bayswater', *Gardener's Magazine*, 7(40), 593 (1832).

Campbell, of Bayswater, was appointed as Curator of the Manchester Botanic Garden, a man who was experienced both botanically and horticulturally with exemplary testimonials.[2] Campbell was first encountered in the minutes on 2 February 1833 providing scientific information, the degree of heat possible from the new boiler in the hothouses.

Campbell had several staff under him, the lodge keepers, the gardeners, and the apprentices. Many of the lodge keepers were women, and their husbands were employed in the garden. The position included a house in the lodge, a wage, coal, and candles. The lodge keepers were required to supervise admissions (checking members, their guests, and strangers), and offer general assistance, especially to the ladies. Checking strangers could be fraught with danger as the minutes recorded on 8 July 1833 that the Rule 'has been in very numerous instances unobserved'; residents of Manchester posing as strangers. On 5 June a member had introduced a Mrs Evans of Birmingham 'who under peculiar circumstances affirmed that name.' Unfortunately she was a resident of Chorlton Row, a short distance from the Garden.[3]

It does seem to have been a difficult job as there was a high turnover of staff before the appointment of Mr and Mrs Ankers. For example, members complained that they were not able to speak to garden staff when they wanted and the lodge keeper was given a bell to ring when a proprietor entered the garden to alert the gardeners.[4] The 1841 census returns for the garden show that a Mr and Mrs Ankers were in place and still employed in 1851. Even though they were long-standing employees, when a deputy Curator, Nickson, was appointed in 1853 (see below), they were told that their lodge was needed for him. Their tenure ended and Mrs Ankers was given a £5 gratuity; she sought further money from the Council, pleading distressed circumstances, but they refused to countenance her request.[5] When Nickson was rehoused, the lodge keeper's position was re-advertised. Perhaps it is not surprising that the Ankers re-applied; it is obvious from her letter that they were in dire financial straits and the job offered a house and income. No doubt pleased not to have to train new staff, the Council re-appointed them, though it would be good to think they felt a moral responsibility to help. The Ankers' duties had changed little, punctuality in ringing the bell, keeping the interior and exterior of the lodges clean, and attending the ladies. They had to provide a security of £50 and either party could give a month's notice.[6]

Campbell, of course, supervised the gardeners. There is no record of how many gardeners were employed when the garden was founded. The 1841 census showed six gardeners were employed, though the number had increased by 1845 when nine were paid as night-watchmen to prevent the theft of lead between December 1844 and April 1845; a sum of £8 was allocated to cover their wages.[7] By 1851 the census shows that they were down to five. Difficult finances appear to have been the explanation for the decrease. The report of the Garden sub-committee in 1851, referring to the management of the new pleasure gardens, was opposed to the suggestion made the previous May by the Finance Committee to save £31 in labourers' wages by dismissing staff; they felt that this could not be done without 'materially neglecting the Gardens'.[8] The Finance Committee apparently won out. Evidence exists that the Council did take care of its staff. Due to a steep rise in food prices by February 1854, the Council increased the wages of the gardening staff, then five labourers and four under-gardeners, by 2s. a week, though they were to be reduced when the Council thought fit.[9]

Gardening work increased as a consequence of the continuing conversion of the Garden to a pleasure ground and an extension to the lake. As a result of all the extra work, in April 1853 Campbell suggested he should have a deputy. On 27 May the Council agreed that they would engage:

> ... a practical gardener at a salary not exceeding 30/- a week to attend to the cultivation and management of the plants in the stoves and Conservatories and the propagation and raising of specimens required in such an institution, such gardener be under the direction of Mr. Campbell and to make himself generally useful in providing plants for the outdoor department.[x]

It was probably no coincidence that Campbell was aged 68 when he requested help. Nickson was appointed to the new post of Campbell's assistant and he and his wife were to live in the lodge. In June 1854 the Council decided an enquiry into the state of the plants in the houses should be undertaken to assess the success of Nickson's appointment.[11] Unfortunately Campbell and Nickson did not get on and by 10 July Campbell had complained to the Council about his deputy.

After a long discussion, not minuted, Nickson and all nine labourers and

Frozen Out Gardeners. From G. Everitt, *English Caricaturists & Graphic Humorists of the Nineteenth Century* (London, 1893).

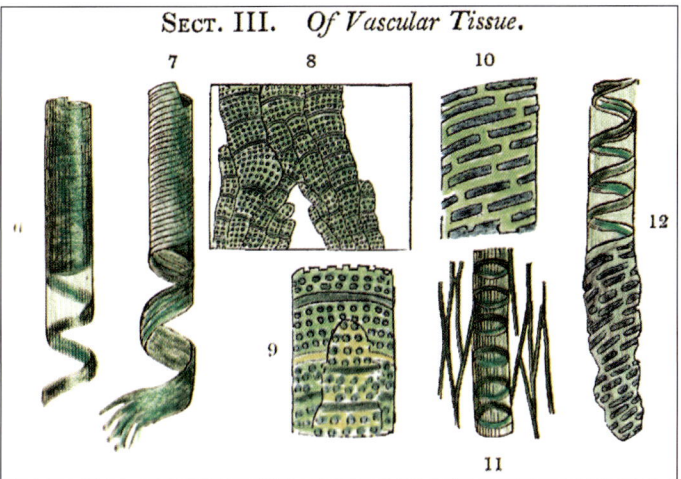

Sect. III. Of Vascular Tissue. From J. Lindley, *An Introduction to Botany* (London, 1832). Lindley was Professor of Botany in the University of London and Assistant Secretary to the Horticultural Society of London at this date.

under-gardeners were discharged; Nickson was re-engaged as foreman under Campbell's supervision.[12]

No reason for this can be deduced from the minutes and it seems contrary to the care for the staff shown the previous February. Over the following months the dismissed men were re-engaged by Campbell.[13]

Campell's final responsibility was the supervision of the apprentices. The records of these apprentices appear irregularly in the minutes; some entries concerned their education, some their performance, and some were of a personal nature. In 1827, the founders of the Botanical Society had proposed to employ apprentices at the Garden, recognising that the current education of gardeners was insufficient for the needs of their situations.[14] The *Address* assured potential subscribers that no-one would be employed at the Botanic Garden who had not supplied testimonials to his diligence, integrity, and skill. Manchester's scheme was innovative. The education of gardeners later became a national concern and in 1840 the Horticultural Society of London published proposals that the regulations used at their Chiswick Garden should be adopted generally for the training of garden apprentices. Campbell was included in a list of advisors consulted by the London Horticultural Society, the only subscription botanic garden curator to be acknowledged.[15]

The Council's stated aim was to produce skilled gardeners knowledgeable in botany and horticulture and who would acquire the habits of self-control, discipline and good conduct, which would enable them to manage other garden staff in future employment.[16]

Notice

The Council of the Manchester Botanical and Horticultural Society announce that they will be at the Gardens on the 3rd. day of May next at 10. O'clock in the forenoon precisely to proceed to the Election of Four Young Men, or APPRENTICES, to be instructed under the curator in Botany and horticulture and to be employed in the Society's Garden.

Manchester 27 January 1830[17]

In February 1830 a sub-committee of six was formed to supervise the Manchester apprentices.[18] Four boys were to be appointed though the Society planned eventually to have 16 apprentices. On 5 April 1830 six boys, together with their parents, were interviewed at the Society's office on Marsden Street, Manchester; their details had been printed and sent to subscribers with a date for a general meeting to elect the successful candidates. The minutes of 30 April record that the four boys started on 9 June. The remuneration of their lodge master was to be 18s a week, an additional £6 *per annum* for instructing the apprentices, and his house would be rent and rates free. The regulations for the master included instructing the boys in botany, horticulture, reading, writing, and arithmetic; any books he wished to use required the Council's approval.[19]

A putter-out supplied the raw materials to hand-loom weavers who worked

St George's Church, Hulme, Near Manchester. From S. Austin, J. Harwood, G. and C. Pyne, *Lancashire Illustrated* (London, 1832).
AUTHOR'S COLLECTION

at home. They were paid on piece rates related to the amount of cloth they produced. They were superceded by the power looms in the textile factories. By 1830 hand-loom weavers were paid minimum wages and in a steep decline.[20] To further the religious education of all the garden staff, the Society had purchased a pew in St George's Church, Hulme.

The sub-committee met at the beginning of June and drew up the rules for the boys' employment. The apprentices were interviewed again at the end of August and they had an oral examination, produced written work and their conduct was scrutinised.[21] The Council agreed that all the boys were satisfactory and indentures were drawn up for their employment. These came into force on 17 November and bound the apprentices for 7 years. The position carried no wages and the boys had to supply most of their own needs: in June 1832 James Harding was supplied with new shoes as the provision of clothing was covered by the Society.

Rules had already been drawn up for the relationship between the boys, the lodge keeper and their medical attendant.[22] The doctor attended monthly and recorded their state of health; the lodge master had been provided with limited medicines. The initial enthusiasm for training apprentices seems to have faded quickly as J.W. Wood, the lodge master, was given one month's notice on 15 February 1832 and the boys were to be supervised by the Curator. The Council approached William Thomson a local lecturer in Botany at the Medical School for his help. He replied on 2 May, 1831:

> To Rev Peter Horden   William Thomson 44 Ossumond Street
>
> The kind offer of the Committee of the Manchester Botanical and Horticultural Society is flattering and highly gratifying to me. … It would be difficult to say in what manner I could make myself useful in the Gardens, except by generally stating that any leisure afternoon work would be gladly devoted to the examination of the young men in such subjects as Smith, Drummond or Russeau, or any other popular authors should have previously made them acquainted with. At the same time I think it right to state to the Committee that the nature of my other engagements would prevent me from giving at present frequent attendances at the Garden, I mean, such as weekly visits. Although I should be able, as often as I can go, to send a considerable time with the young men.

Thomson resigned 10 April 1832. His advertisement on 21 July 1832 in the *Manchester Times* showed that he must have been still on good terms with the Council as he is offering botany classes to lady subscribers so that they 'may have an opportunity of pursuing the study with Mr. Thompson's assistance in the garden. – Terms One Guinea and a Half.'

When Elizabeth Bottril was appointed lodge keeper in February 1834, her duties included caring for the apprentices with an allowance of 9s a week each. This covered board, washing, and cooking, as they lived in the attics at the lodge. The apprentices then disappear from the records and the conclusion must be that the scheme foundered, perhaps as a result of the financial difficulties of the Society. For example, when Wood, the dismissed lodge master asked for a

The Candidates for Apprenticeship

John Bailey	15	Works in kitchen Garden for J. Clegg.	Putter out 7 children	Promising lad for his years
James Harding	16	In Mr. Darbishire's Garden previously	No father Mother 4 children	Fine promising lad in every respect very suitable. A very steady and industrious lad.
Thomas Godden	16	In his farther's garden since he could work	Small farmer and gardener 9 children	Healthy in appearance but not much apparent activity.
Jonth. Morrell	18	Gardener to Mr. Forth then with Mr. Starkie now with S. Hooton	Gardener to Mr. Starkie Hunt Royd	Promising intelligent young man with some experience of plants
James Dean	19	With John Holland Chorlton, 2 years	Gardener to S. Philips 5 Children	Healthy young man, too old
William Moore	16	With Mr. Hunter but not as gardener	Joiner 7 children	Likely in appearance to suit but at present ignorant of gardening

Though four boys were appointed as apprentices for a trial period on 30 April 1830, their names were not entered as appointees in the minutes. One was John Bailey, recorded as given permission to visit friends on a Sunday but to return by the evening.[1] James Harding was another who was called before the Council for a misdemeanour on 7 September 1831. They put him on trial for three months doing spade labour only, his behaviour was monitored and, happily, he survived. Jonathon Morrell was the third. The other apprentice's name has not been found.

Minutes, 30 June 1830, MBH 2/1/1.

character reference, the Council agreed he was no trouble but stated that the Society could not afford his services.[23]

What work the apprentices did is not recorded though on 24 November 1830 the minutes note they were set to letter the tallies. Later they had to paint the extremities with red lead to stop them rotting. One apprentice stands out from the records; James Morrell was accused of leaving the garden 'with property belonging to the Society'. The Council sent a letter to his father explaining in view of his offence a warrant would be issued for his arrest. A letter from his distraught mother is attached to the minute book:

Rules Governing the Apprentices

1. That a book be procured in which the conduct of each Apprentice every day shall be noted by the Lodge Keeper under the heads of Very Good – Good – Careless – Idle – Bad, and that the entry shall be publicly read before the Apprentice separate. He had to make regular reports to the Council.
2. That a copy of Pinnock's Catechism be procured for each Apprentice and one for the Curator, for the purpose of being committed to memory, and 5 copies of Abercrombie's Gardener for general reading.
3. That books be made of plain paper containing ¾ quire and stitches and covered with pasteboard in thick glazed paper, in which each apprentice shall enter his daily occupations.
4. That Wilkinson's Arithmetic be ordered.
5. That 5 slates, 5 plumb ruling pencils, three inkstands and pens and pencils for slates be ordered.

The lodge master was required to see them sign in and out of the gardens each day, accompany them to church on Sunday, and every evening the apprentices were to take part in the prescribed day's activities under his direction. These were laid out as: Monday and Tuesday, instruction in Botany and Horticulture and written work on the subjects; Wednesday register only; Thursday study Arithmetic and Writing; Friday the same as Monday and Tuesday; Saturday the same as Wednesday. Their conduct during these hours was to be noted by the lodge master in the book provided.

Minutes, 11 June 1830, MBH 2/1/1.

Frontispiece from T. Mawe and J. Abercrombie, *Every Man His Own Gardener* (London, 1813).

Budding's Mowing Machine. From J. C. Loudon, *An Encyclopaedia of Gardening New Edition* (London, 1834). The Minutes, 4 August 1831, noted that a grass cutting machine was needed and the Curator was to try out Mr Lot's machine at his garden in Regent Road, Salford, as it supposedly gave superior mowing. This must have been the mechanical lawnmower patented by Edward Budding in 1830. Mowbray was later authorised to buy one for use in the garden. AUTHOR'S COLLECTION

We were thunderstruck when his brother came and told us he was left. I am almost distracted, and his father, poor man, is grieved beyond Everything, for we have been lifted up, knowing he was in a good place.

Out of respect for the parents, the Council decided to merely dismiss him with a severe reprimand. To understand the parents' distress one only has to consider a report read before the Manchester Statistical Society on 6 January 1836. The figures showed that in the Boroughs of Manchester and Salford 'at least 20,000 children between five and fifteen are under no course of instruction'. An apprenticeship at the Botanical Garden must have seemed a salvation to many hopeful families. The apprentice scheme seems to have been abandoned by the 1840s probably due to financial considerations.

Though the gardeners, of whatever rank, are mainly absent from the official records they were the men who laboured behind the scenes in the years that follow and deserve to be remembered.

Notes

1   J. C. Loudon, *New Encyclopaedia of Gardening New Edition*, §7040 (London, 1834).

2   The Birmingham Society appointed David Cameron as Curator to the Birmingham Botanic Garden in 1831. Application and testimonials for that post are extant. See Appointment of Curator, M/S 1520/33, Archives, Birmingham Botanical and Horticultural Society, Birmingham Central Library, Birmingham.

3   In London ladies of the demi-monde proved a problem inpublic. 'These demi-reps make peremptory conditions that they shall have broughams for the Park and tickets for the Horticultural and even for Fêtes the Botanic Gardens.' (Quoted in C. Pearl, *The Girl with the Swansdown Seat*, 99 (London, 1955).

4   Minutes, 22 September 1830, MBH 2/1/1.

5   Minutes, 15 April 1853, MBH 2/1/3.

6   Minutes, 17 April 1854, MBH 2/1/3.

7   Minutes, 4 June 1845, MBH 2/1/2. They were listed as Thomas Campbell (30), Thomas Forbes (30), George Adams (29), Lot Holt (29), Thomas Turner (29), Christopher Priestman (26), Thomas Cheetham (26), John Deas (18), James Kidd (30), David Pendleton (gone to Liverpool), Thomas Greville (gone to E. Loyd).

8   'Report of the Garden subcommittee', Minutes, 23 October 1849, MBH 2/1/2.

9   Minutes, 3 February 1854, MBH 2/1/3. The labourers were Thomas Cheetham and Thomas Turner, both 15 years service, Jonathon Hampson, 9 years, John Hulme, 4 years and William Hampson, 24 years. Each earned 12s. a week. The undergardeners were Thomas Campbell on 18s. a week and William Curren, William Nutter and Samuel Norris each on 13s. a week (Thomas Campbell had appeared in the 1841 census as Alexander Campbell's son aged 15).

10   Minutes, 8 April 1853, MBH 2/1/3. The advertisement they were to place requested 'A married man without family would be preferred. Wages were 30s. a week with a residence on the premises.

11   Minutes, 22 June 1854, MBH 2/1/3.

12   Minutes, 10 July 1854, MBH 2/1/3.

13   Minutes,13 April 1855, MBH 2/1/3.

14   *An Address to the Inhabitants of Manchester and the Neighbourhood on the formation of a Botanical and Horticultural Garden*, 2 (Manchester, 1827) (hereafter *An Address*); EGR 4/2/10/20/1, Dunham Massey Archive, John Rylands Library, Manchester (hereafter EGR).

15   G. Bentham, 'The instruction of young men in the art of gardening', *Transactions of the Horticultural Society of London*, section 8, vol. 2, 2nd series, 454–8 (London, 1842).

16   See also 'Of Superintendance and Management of Gardens': J. C. Loudon, *Encyclopaedia*, §§3164–315 (1834). His book *Self-instruction for Young Gardeners*, was published posthumously in 1845.

17   Minutes, 25 January 1830, MBH 2/1/2.

18   Minutes, 1 February 1830, MBH 2/1/1.

19   Minutes, 9 June 1830, MBH 2/1/1.

20   Minutes, 30 June 1830, MBH 2/1/1.

21   Minutes, 1 September 1830, MBH 2/1/1.

22   Minutes, 30 November 1830, MBH 2/1/1.

23   Minutes, 1 March 1832, MBH 2/1/1.

CHAPTER EIGHT

'It is not a church, not a palazzo, not a chateau'[1] 1839–1854

....

'The Gardens are very tastefully laid out in the ornamental style. There is an extensive arboretum, containing some fine specimens of various kinds of trees and shrubs, planted on the east, west and south sides of the Gardens; such kinds as are most conspicuous and interesting, being placed in more public situations.'

Benjamin Love, *Manchester as it is* (Manchester, 1839)

By 1839 the original planting was maturing and creating the ambience envisaged by the founding members. The wall and the trees would now have screened the Garden and walking within allowed subscribers to stroll in their own '*hortus conclusus*' without the prying eyes of the industrial masses who were only allowed into the garden during the summer exhibitions. Love's article further describes how by taking advantage of the natural springs in the lower garden the extensive lake was 'highly picturesque' and crossed by a 'romantic-looking bridge' (erected in 1835). Crossing the bridge brought the visitor to 'an Alpine region of miniature rocks and caverns' and a rockery of Derbyshire tufa – still

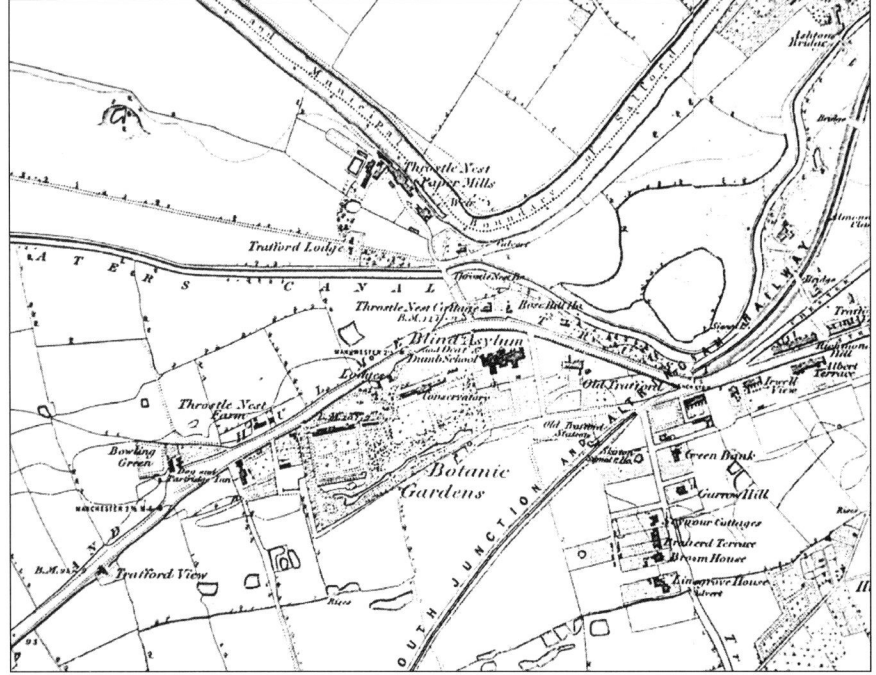

The Manchester Botanic Garden 1848, prior to the alterations. From Ordnance Survey, *Manchester and Salford* (London, 1848).
TRAFFORD LOCAL STUDIES, SALE

An Aside: Finances 1832–1845

What is remarkable throughout the Society's history is that the ongoing costs of maintaining the gardens, let alone the cost of new builds, rarely seem to have been budgeted for annually in advance. Bank loans or mortgages were raised retrospectively to cover debts created by major projects with the increased value of the land being used as collateral. In January 1833 to reduce costs, Mr Walker visited the Warden of the Collegiate Church who agreed to relinquish their claim for tithes on condition the ground remained a botanic garden. One major saving was plant donations, which were numerous throughout the Society's history. Though in May 1833, when a sub-committee was appointed to address the need for additional funds, the Council declined to purchase a collection of plants due 'to the state of the Society'. The major annual event, and financial outlay, was the opening of the Garden during Whit-week when exhibitions were organised and the public admitted for a fee. In 1839, Love, referred to the sorry finances of the Society in *Manchester as it is*:

> The income of the Botanical Society for the years 1837–8 amounted to £1,443. The expenditure, owing to an extraordinary outlay, was £2,185. Of this sum £1,013 had been paid on account of the new conservatory and forcing houses.[1]

The annual *Report* for the year included a mortgage of £5000 with the Manchester Assurance Society against which interest payments of £212 10s were made. The Council was sanguine about the debt as, they reminded subscribers, the value of the land and property had increased. By 1841 there was no improvement and the annual interest payment was £362 15s 10d. In spite of this, a special general meeting of hereditary members still voted to increase the mortgage by £800 at £5% per annum. The main on-going cost during the period was the need to maintain the fabric of the Garden. The changes of 1848 and the new exhibition House in 1853 were to challenge finances still further.

1   Benjamin Love, *Manchester as it is* (Manchester, 1839), 121.

the classic rock for growing alpine plants today. The Garden praised by Joseph Paxton in 1831, had realised the subscribers' dreams.

The Council continued to add attractions for members. Garden seats were installed and in 1836 a glasshouse was built to grow both hothouse and cool greenhouse plants for sale to members with a scale of charges drawn up. A members' request book was started for inclusion in the annual seed distribution, and members came to tasting days of the apples and pears from the fruit garden. Grafted scions, prepared by Campbell, could then be purchased of their favourite fruit. A Mr Silvester of Chorley offered the Society several tropical aquatic plants, if they could be housed correctly. The Council knew this would be a great draw to new subscribers if they could grow the Amazonium water lily, the *Victoria Regia*, a sensation when it was introduced to Britain. They were quick to make alterations to existing facilities to house the plants. The Eastern portion of a new house containing a pineapple pit was converted and pans and tanks ordered to accommodate the aquatics. In the gardening world, Manchester was noted for cultivating orchids. Several subscribers had national

Rustic bridge at Tatton Park, Cheshire. The bridge is similar to one from the Manchester Botanic Garden photographed in 1857 and illustrated (in a poor reproduction) by H. Gartside in *The Royal Jubilee Exhibition, Manchester, 1887: a photographic record* (Manchester, 1877).

reputations in the field and had had varieties named in their honour. Perhaps it was they who suggested that the remaining pine pit was halved so orchids could also be grown; boxes of orchidaceous plants are listed in the gifts to the garden. All this at a cost of £20!

Other costs began to rise. Repairs to the original building were beginning to be noted in the minutes; trellises needed patching, repainting was required, the Vinery had loose flags and decaying woodwork. In June 1845, £100 was paid for 50 extra garden chairs and there were damaged chairs needing restoration. 1845 saw plans to build a new house for the Herbarium that did not come to fruition due to the cost. May 1846 did see the purchase of a new canvas top for the Exhibition tent, though Campbell was instructed to spend no more than £28. Lead was stolen from the roofs and extra wages were paid to the gardeners for forming a night watch. With finances difficult and many buildings in a bad state of repair, membership of the Garden was becoming less attractive.

Help could have been at hand for the Society. In 1846, the introduction in Manchester of a Saturday half-holiday offered an opportunity to improve the finances by opening the Garden to the public. A contemporary description of Manchester makes clear that walks in open spaces would have been a popular occupation:

'As regards the town itself, it is a mass of dirt and confusion, smoke enters your mouth as you talk, into your eye as if you look upward to observe if there are any symptoms of dry weather and covers your face with anything but flakes of snowy whiteness'.

Manchester businesses (many run by the members) responded enthusiastically:

'... that the experiment on a large scale has been made and it has succeeded. In Manchester ... the city which is devoted to business, most of the merchants and wholesale shopkeepers have for some time past closed their counting houses and shops, by 2 p.m. on Saturday afternoon. All the clerks, workmen, and apprentices, are thus set free for a half days rest and recreation'.[2]

The *Victoria Regia* is still popular today (Edinburgh Botanic Garden 2005).

The Society however still targeted their appeal to the upper classes: 'To the wealthy inhabitants of our densely populated town, the opportunities which these beautiful Gardens afford for needful recreation in a purer atmosphere can hardly be overrated'.[3]

Manchester publishers produced guidebooks giving walks and railway excursions including the area near the Botanic Garden, though the Garden itself was not included.[4]

'There are several beautiful walks ... in MOSS SIDE from Stretford New Road, and as far as Chorlton-cum-Hardy, quiet and rural, although on a clear day the Derbyshire Hills form a beautiful termination to the prospect'.

An Aside: The Competitors

By 1846 competition to the Botanic Garden came from the Parks and from commercial organisations. The Zoological and Horticultural Gardens at Belle Vue, East Manchester, had opened in May 1837 and offered spectacle within a pleasure garden. A second public pleasure garden was operating much nearer the Botanic Garden, Pomona Gardens, established in May 1846 on land between the Irwell and the Bridgewater Canal. Both offered seven days a week opening and cheap admission: between 6d and 1s, with children half-price. Local advertisements show they put on spectacular displays, especially at the Whit-week holidays; in 1850 Pomona staged the eruption of Vesuvius and in 1852 Belle Vue the shelling of Algiers. The Council did not consider these gardens as competition as the exclusivity offered at the Garden attracted a different clientele. However the competition from the Park Movement was attracting members away from the Society. How far they were a real threat was demonstrated by this extract from *The Manchester Literary Times*, 3 June 1848; 'The people's parks are crowded. No more beautiful scenes could be desired by man than such as are presented on the holiday time of Saturday in any one of these cheerful places of resort.'

In May 1849, after much heated discussion, public admission was allowed on Saturday afternoons for 3d and the names of visitors were to be entered in a book.[1] The Council implemented other changes, refreshments were available on Promenade and Exhibition days and Campbell constructed a labyrinth as a new attraction. Though the wealthy were unlikely to desert the exclusive grounds, the Society relied heavily on the membership of the middling ranks of the middle class who might prefer the parks.

Source: Minutes of the Manchester Botanical and Horticultural Society except where stated.

The Italian Garden. *Zoological Gardens, Belle Vue, Manchester* (n.d.).
AUTHOR'S COLLECTION

1   Birmingham had already allowed non-subscribers into the Garden and the minutes of their 1844 annual meeting recorded that the working classes were allowed in on Monday and Tuesday for one penny, although they were excluded from the hothouses, not allowed to picnic or smoke and two policemen were employed in case of trouble.

By 1856 when a national half-holiday was proposed, Manchester was cited as the example to follow. Lord Shaftesbury declared, 'Go now to the City of Manchester, and see that ... which in former times were hives of turbulence, of dissipation, and of peril to the community and you will see an orderly, and decent ... a most highly educated population'.[5] A letter from Archibald Prentice the same year mentioned that the Manchester Botanic Garden was open to visitors on Saturdays:

Barlow Hall, Chorlton-cum-Hardy, residence of member William Cunliffe-Brooks, Banker. From L. H. Grindon, *Country Rambles and Manchester Walks and Wild Flowers* (Manchester, 1858). This work also contains his accounts of trips with the Manchester Field Naturalists Society. 'There are several beautiful walks ... in MOSS SIDE from Stretford New Road, and as far as Chorlton-cum-Hardy, quiet and rural, although on a clear day the Derbyshire Hills form a beautiful termination to the prospect'.
AUTHOR'S COLLECTION

'In addition to these beautiful places of resort [the Parks], at certain seasons, the Botanical Gardens, the Natural History Society's Museum, and the picture gallery of the Royal Institution are thrown open to Saturday visitors at a merely nominal rate of admission'.[6]

In fact, the pressure on the Society from these changes in public habits had become clear as early as 1846 when the Council admitted that there was nowhere to shelter or promenade in the rain at the Garden, a facility offered by the new Parks – the new tent roof had been the answer then.

This was not the answer now and the subscribers met and demanded that the Council make dramatic changes. The enlightened non-conformist ideals of science and horticulture no longer attracted subscribers. The current membership were interested in the Garden becoming a pleasure ground, nothing more. The Council conceded and in 1848 science was eliminated and the horticultural garden and its heated walls and greenhouses were dismantled.[7]

The glass covering the peach wall was used to construct a propagating house. Two hundred and twenty fruit trees were named, priced and advertised for sale. Promenade evenings and Exhibition days were to include refreshments and a printed list of the music to be played was sent to members.[8]

Though these changes were made, by 1849 the finances had not improved. The changes did not bring about the desired increase in membership. Subscriptions were not being paid on time or renewed; even a reduction in the subscription had not helped. The extent of the problem may be judged by the fact that nine members were sued for debts owing for produce from the Garden. By August the Society 'named and shamed' defaulting members, and threatened them with proceedings. The Garden and Finance sub-committees presented reports in October 1849. Both make salutary financial reading. The finances showed that, even if the expenses were modest and 'rigid economy … practised', the debt would increase. (They did not take into account the mortgage interest or the principal borrowings.) Finally they stated 'the gardens cannot be worked with pleasure, or be kept in [a] state of efficiency', an admission of the failure of the Society's new policies. The Garden report was equally depressing. It opposed a saving in labourers' wages; this could not be done without neglecting the new Garden as this required additional labour, not less. They concluded that there was 'the want of proper funds to keep a collection of plants worthy of the purposes for which the Society was established'.

In 1850 the Council feared the Garden would have to be discontinued the following year and regretted it was 'discreditable' that 'the second city of the British Dominions should, for want of adequate pecuniary support, be forced to abandon gardens most tastefully laid out.' Yet in the next paragraph they list the new attractions under construction: the new pleasure ground, the extension of the lake, the labyrinth, an island approached by two new rustic bridges and a new mound, topped by an arbor. At a special meeting in May 1851, the Council reported that the above work was in hand and that by 'the liberality of their Bankers all the debts have been paid off'. In fact, to pay off pressing debts, an additional £300 had been borrowed and the Council was considering selling the land. The only hope was that by dividing each hereditary share into three, thus reducing the cost of a share to one guinea (£1 1s), many more subscribers would join and the future would be secured.

Many of the buildings were in a dilapidated state. With little money available only the orchid house was repaired and by the end of 1850 all the principal ranges needed attention. It would seem that essential repairs were only carried out *in extremis*. In April 1851 Campbell was instructed to repair the doors to the conservatory, the lights in the lean-to houses, the roof over the lodge keeper's kitchen, and the bridges. However 373 new members were recruited in 1851.[9] Perhaps the increase in membership encouraged the Council to embark on another expensive construction, a new exhibition house; the previous minutes make clear that, even with a new covering, the exhibition tent was decrepit and offered little protection from the weather. Members were requested to make donations but only £1086 2s was forthcoming, insufficient to cover the cost.

One of Manchester's first parks, Queen's Park, formerly known as Hendham Hall, situated near Middleton to the north of Manchester, was acquired in 1846. The grounds had been laid out the previous year by Joshua Major, a noted landscape designer and author.

Nevertheless in September 1851 the Council went ahead and considered plans and specifications and called for further donations. On 8 March 1852 Campbell was asked to stake out the new building in line with the main gravel walk and the conservatory. The building was to be 200ft long, 60ft wide and 20ft high (c.61 × 18 × 6m), and a local firm, Fox and Henderson, were asked for a design and estimates. By May their estimates were deemed too expensive and at the same time money was released to repair the water-lily house; a new one was too expensive so the old one was reglazed. It was still make do and mend. Plans for an Exhibition House were still being examined though money was still not being subscribed in sufficient amounts. A member of the Council, the Revd Gibson, personally contacted members to explain the need for the Exhibition House and request subscriptions; his canvas of members secured the promise of £800. At a special general meeting on 10 September, it was decided to raise the estimated cost of £2020 by donations. A Mr Risley's plans were recommended for acceptance. The estimate was £3515, with additional costs for wrought ironwork of £2050 (or with ornamental extras, £2390), and window glazing was an extra £822–£900.[10] Undaunted, the Council asked for a second estimate in October from the Britannia Foundry, Derby, which came in at £7250 not including glazing, painting, foundations, building and joinery. The minutes of 4 November note that the Exhibition House project had been abandoned. In December the finances were such that the Treasurer overdrew by £350 to pay outstanding debts.

The project must have been resurrected in the winter, as by the annual meeting of March 1853 a Manchester architect, Thomas Worthington, had been found to design the House. He was given a budget of £2500 and asked to prepare estimates for which his remuneration would be £100.[11] An Exhibition House sub-committee met in May and approved the agreement with the builders, Bellhouse and Son (several of the family were members), and the plans were exhibited at the Whit Week Exhibition (Unfortunately not found).[12]

The Whit Week Exhibition for 1853 was to be the Society's Jubilee Exhibition and the Council decided special efforts should be made. This could only mean more expense. Sir Joseph Paxton, Sir William Hooker, the Duchess of Northumberland and Robert Marnock were requested to send special plants. To advertise the show, placards were ordered for the local railway stations and for omnibuses going in the direction of the Garden. Unfortunately, as the weather proved very bad, the exhibition was a financial disaster. The Council drew the conclusion that an Exhibition House would have saved the day and they pressed ahead with the project though adjourning further decisions until an estimate for the new building was presented. At the annual meeting in March 1854, the Council expressed regret that they had not been able to erect the Exhibition House in 1853 due to lack of funds though they claimed they had £1070 in the bank and a loan of £500, promised in case of a shortfall, from one of the members, Edmund Buckley.

This would make little inroad into the debt, especially as the Council pressed on with the new Exhibition house. By 4 August 1854, the Exhibition House was in the process of erection and £1000 had been paid to Bellhouse, the builders, on account. The Council agreed to allow the plans to be exhibited by Worthington (the architect) at an exhibition at the Royal Institution, perhaps hoping they would attract new members. These were desperately needed as the same minutes record a demand from the Manchester Insurance Company requiring the £5000 mortgage to be redeemed within six months.

The Exhibition sub-committee met again on 24 August and announced that 'the following noblemen and gentlemen be the patrons of the ensuing Exhibition of the New Exhibition House'. There followed 36 names of the great and the good of Manchester including the Earls of Ellesmere, Wilton, and Stamford and Warrington, the Lord Bishop of Manchester, Sir Humphrey de Trafford and Wilbraham Egerton, Bart. The opening, to take place on 12 September, was to include:

> 'plants, flowers, fruits, vegetables, artificial flowers and fruits, vases, flower pots, baskets, garden sheds, seats, iron works, garden implements, which may be exhibited from all parts of the Kingdom free of charge'.[13]

The materials for finishing the Exhibition House were ordered together with dies for medals and bands were engaged for the four days of the Exhibition. September was less than a month away and the building was far from completed. In fact there are no reports in the minutes or the press that the Exhibition ever took place.

The Exhibition House was finally completed and it is surprising no report appeared as Worthington had produced a spectacular design for the Society. At the beginning of October, Worthington met with the Committee to inspect the House, report on any outstanding work and the likely time to completion. Disaster struck, the roof sheets of the new building had sunk. At a special meeting in late October Worthington recommended that cast iron purlins needed to be

The New Exhibition House. 'It is very original. For a while we trace the ecclesiastical in the general plan ... and the style of the architecture prevalent in Italy or Switzerland. It is not a church, not a palazzo, not a chateau ... here an arch or a wheel window, there a square campanile ... all judiciously introduced and blended'. *The Builder*, 6 November 1854, 11. The Exhibition House was 150ft long by 57ft (*c.*46 × 17.4m) wide with a central nave supported by cast iron pillars. The space beneath one of the wheel windows was to be used by an Orchestra. The glass used was the most up-to-date available, Hartley's patent for the roof and Chance's patent for the upright portions. On the south side the Council planned to build a wide terrace. Cecil Stewart in *The Stones of Manchester* claimed that the design was 'the most remarkable product' of Worthington's European tour. Worthington was still under thirty at this date and Stewart called it 'an indiscriminate sowing of wild oats'.

inserted to support the roof and this was done at once. The Treasurer, Birmingham, died in February 1855 and John Butterworth, his successor, was asked to settle Worthington's bill as he was demanding payment even though the new building was still causing problems and spouts had to be fitted as soon as possible to carry water from the upper roof to prevent flooding. The accounts delivered to the Council the previous September showed that cash in hand amounted to £59 12s 6d and that the new building fund contained £165 10s 10d after the payment to Bellhouse. In addition, Bellhouse had tendered for twenty tables at a cost of £2 each, which was accepted, thus leaving £19 12s 6d in the Exhibition Fund. Tenders were also invited for painting the house and 300 yards (*c.*91.5m) of green cloth were bought for the displays. A member, Mr Rothwell, a painter and decorator, painted the Exhibition House *gratis* in March.

Presented at the Annual Meeting on 5 March 1855, the *Report* covering 1854 stated that 'in consequence of the Erection of the Exhibition House ... a very considerable fresh debt has been unavoidably incurred'. Seven hundred and forty eight proprietors had made no contribution to the fund and they were reminded that if they would do so all would be well; only six guineas was forthcoming. The meeting noted that £1367 17s 11d was still owing for the new building, and Bellhouse the builder and Worthington the architect, were pressing for payment. It seems incredible, given the situation, that this *Report* also records that further alterations to the grounds had been made in preparation for Whit week: the tent had finally been disposed of, new walks had been made, trees had been planted along the North wall for extra privacy and a terrace 30ft (*c.*9m) wide was being constructed along the south front of the Exhibition House together with a water feature. In fact it is hard to understand the confidence with which the Council predicted the future success of the Society, especially

Cunliffe, Brooks and Co's Bank, Market Street. From S. Austin, J. Harwood, G and C. Pine, *Lancashire Illustrated* (London, 1832). Cunliffe, Brooks and Co.'s bank had premises at 75 Market Street. Samuel Brooks was a founder member living in the town centre at 12 Moseley Street, Manchester. His son William was also a member and lived out of the city centre at Barlow Hall, Chorlton-cum-Hardy (see above). Many families maintained their connections over the generations
AUTHOR'S COLLECTION

as the minutes of 13 April 1855 recorded that the Treasurer and the Chairman, Michael Potter, needed to borrow £1000 'from any source'.

The minutes are then missing until 1858; the *Report* for the year 1855 (presented on 3 March 1856) was reported in the *Guardian*. This informed members that £1000 had been borrowed from the Lancashire Insurance Company. This was probably the £1000 that the Treasurer and Potter had been asked to borrow from 'any source' and, together with the money from Edmund Buckley, was used to settle the Exhibition House account; the Lancashire Insurance Company had already loaned £5000 on mortgage in November 1854. The Council regarded the Society as prosperous, claiming the increasing subscription income and the £1000 extra loan, would be available to the Society for 1855, although they admitted this would not cover all the outstanding expenses. The structures in the Garden were reportedly in excellent condition as expenditure for the previous year had included new water tanks in several glasshouses for economy of watering and the installation of new boilers. Surprisingly given the need for income, one source, public admission on Saturday afternoons, had been discontinued. Since 1849, the 3d public admission on Saturday afternoons, which raised £43 17s that year, had by 1855 risen to £105 14s, 5% of the Society's income. Given the financial situation, it might have been prudent to adopt the practice of the Birmingham Botanical Garden and open the Garden to the general public more frequently. Not for Manchester! Despite the rise in debt, the *Report* noted that public entry was to be discontinued in the pursuit of exclusivity, for the Council could not 'recommend a continuance of this practice, as they find it is strongly objected to by many members'.

Notes

1   *The Builder*, 6 November 1854, 11.

2   Anon., *Manchester and the Manchester People with a Sketch of Bolton, Stockport, Ashton, Rochdale and Oldham and their Inhabitants by a Citizen of the World*, 13 (Manchester, 1845).

3   J. Fitzgerald, *The Duty of Procuring more Rest for the Labouring Classes; the Earlier Closing of Shops and the Saturday Half Holidays*, 66 (London, 1856).

4   *Report of the Council of the Manchester Botanical and Horticultural Society* (Manchester, 1847), MBH 7/3/3.

5   *Half-Holiday Speeches, Exeter Hall Meeting, April 24, 1856*, 5 (London, 1856).

6   Quoted in J. Lilwall, *The Half Holiday Question Considered with Some Thoughts on the Instructive and Healthful Recreations of the Industrial Classes*, 2nd ed. (London, 1856).

7   Minutes, 24 January 1848, MBH 2/1/2.

8   The band of the regiment stationed in Manchester played at events. The officers, and their families were given free admission to the Garden by presenting their visiting card.

9   Unusually the minutes for the year 1851 listed the names of new members recruited each month.

10   Minutes, 2 September 1853, MBH 2/1/2.

11   Worthington became one of Manchester's most famous architects. See A. J. Pass, *Thomas Worthington* (Manchester, 1988).

12   See D. R. Bellhouse, *David Bellhouse and Sons: a Manchester building business* (London, 1992).

13   For a description of flowers shows see M. R. Blacker, *Flora Domestica: a history of flower arranging 1500–1930*, 176–8 (London, 2000).

CHAPTER NINE

'Art Triumphed over Nature' 1857

Art-Treasures Exhibition Building 1857. The Art-Treasures Exhibition Building looking north-west towards the Botanic Garden. From H. Garside, *The Royal Jubilee Exhibition, Manchester, 1887: a photographic record* (Manchester, 1877).
CHETHAM'S LIBRARY, MANCHESTER

The year 1857 was a remarkable one for Manchester. Axon recalled that 'The great event of the year was the Exhibition of the Art-Treasures of the United Kingdom, at Old Trafford, which demonstrated the wealth of the British artistic possessions'.[1] The Exhibition site was on a plot of land adjacent to the south side of the Botanic Garden. To complement the Art-Treasures Exhibition the Council of the Botanical and Society arranged a series of special exhibitions. The correspondent of the *Courier* reported on 31 May:

> The Magnificent collection of American plants, occupying the gigantic tent erected purposely for their display ... are now in almost full bloom, and present one of the most beautiful sights ever seen in these gardens.[2]

Their correspondent had reported the previous week, 'Mr. Campbell, the able curator of the gardens, deserves unqualified praise, for his great exertions to improve the Society's property, and please the public'. At a Special Meeting of Proprietors on 26 September 1857 the members learned the true cost of the Art-Treasures Exhibition to the Botanical Society. The event, thought to be the saviour of the Society's financial distress in 1856, had been a financial disaster.

> They all knew that there was a deficiency in the accounts to a considerable amount. ... a deficiency of £1,000 to be added to the previous deficiency of £1,300. – (Oh!) ... Something should have been done ... to clear off the debt ... It was thought the Art-Treasures Exhibition would have this effect.

What was the Manchester Art-Treasures Exhibition, and why did the Society expect it to salvage their finances?

The suggestion for an Art-Treasures Exhibition was made in March 1856 by several of Manchester's wealthy elite, who had been impressed by the Paris Exhibition of 1855. They also recalled the Crystal Palace Exhibition of 1851 and the Dublin Exhibition of 1853. Their idea was to bring together under one roof an elite collection of art for the 'edification of their fellow men' and published the idea as a paper co-authored by Mr J. C. Deane, Dublin, and Mr Peter Cunningham, London. The proposal was for a building that would hold at least 30,000 people, and could be sold at a good price after the Exhibition. The total expenditure was expected to be £35,000 and the dates suggested were 1 May–1 October 1857. Season tickets would be available; the first day and the Queen's visit (if arranged) would only be open to such purchasers and this would guarantee extensive sales; though they acknowledged that 'to the shillings we must look mainly to success'. Their first suggestion was that a deputation should present the proposal to Prince Albert to seek his approval and co-operation. This duly took place on 7 May where His Highness, in a letter to the Exhibition's President Lord Ellesmere, promised to give his support and gave advice on persuading the owners of important works to consent to their display:

> A person who would not otherwise be inclined to part with a picture would probably shrink from refusing it if he knew that his doing so tended to mar the realisation of a great national object.[3]

On 20 May, the Queen also granted her patronage in a letter received by R. N. Philips, High Sheriff of Lancashire.

On 2 June, when the Art-Treasures Executive Committee announced the site for the Exhibition, its neighbour on the northern boundary was to be the Manchester Botanic Garden. The site was a plot of land covering 17½ acres (c.7 ha) leased from the Trafford Estate for two years and formerly occupied by the Manchester Cricket Club.

A study of garden history shows that in 1857 gardening itself was regarded as an artistic activity and gardens were the subject of many canvases.[4] Given this perception of gardening by the subscribers, the juxtaposition of the Garden and the Art-Treasures Exhibition was seen as a natural alliance and the two committees agreed to cooperate. The Society had its own ulterior motive; partnership between the art of man and the art of nature was to solve their liquidity problems.

However there were more connections than just the site. The proposal circular, dated 22 March 1856, was signed by seven men and four were Botanical Society members: Thomas Bazley, J. A. Turner, William Entwistle and Thomas Ashton.[5] The deputation that waited on Prince Albert at Buckingham Palace consisted of James Watts (Mayor of Manchester), R. N. Philips (High Sheriff of Lancashire), Sir John Potter, Stephen Heelis (Mayor of Salford), Thomas Bazely, Joseph Heron (Town Clerk of Manchester), and Thomas Fairbairn: Watts, Philips, Potter, Heelis, and Bazely were all Society members.[6] A general council was then

Sculpture Gallery,
Art-Treasures Palace,
Manchester 1857. 'The
Art-Treasures Exhibition
is… the most interesting
feature in the history
of the important
and wealthy city of
Manchester. … Now
she steps forward in
her aggregate character
to emulate the glorious
example of Florence
of old, to display to
the world the richest
collection of works of
fine art the resources of
the country afford.' *The
Art-Treasures Examiner
A Pictorial, Critical, and
Historical Record of the
Art-Treasures Exhibition,
at Manchester in 1857*
(Manchester, 1857).
AUTHOR'S COLLECTION

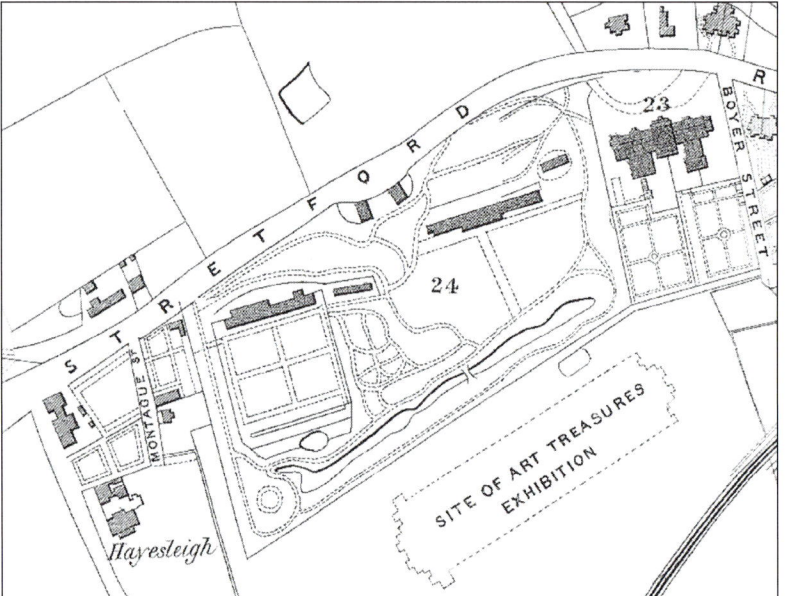

Simm's *Map of
Manchester and its
Environs* (Manchester,
1858). The photograph
of the Art-Treasures
Building (p. 78) was
probably taken from
a garden on the right
immediately below the
railway line which looks
north-west towards the
Garden.
AUTHOR'S COLLECTION

formed to organise the Exhibition, a Patron of the Botanical Society, the Earl of Ellesmere, agreed to be President and again other members were included: these were Watts, Ashton, and Entwistle with Watts being Treasurer and *ex officio* Chairman.[7] Within weeks the sum of £74,000 had been guaranteed to the enterprise and Watts had negotiated a £25,000 loan from the Manchester branch of the Bank of England.[8] The speed and success of the fund-raising for the Art-Treasures Exhibition contrasts sharply with the dismal failures that usually met similar appeals from the same men to the same wealthy subscribers as members

Jean van Huysum,
born Amsterdam 1682.
Manchester Art-Treasures
Exhibition, Saloon
H, Nos 1009, 1010,
Property R. S. Holford
Art-Treasures Examiner.
Robert Steynor Holford
of Westonbirt was a
Gloucestershire, MP
whose family wealth
came from supplying
fresh water to London.
A wealthy man he was
passionate about art and
landscape gardening.
AUTHOR'S COLLECTION

of the Botanical Society. The difference was surely that with the Art-Treasures Exhibition, Manchester's pride in its city and business acumen was to be displayed on the national stage and dispel the city's image as filthy, congested industrial 'Cottonopolis'. Understandably the upper middle class sponsors prized this visible artistic display, rather than the Garden that was becoming less significant to them as they moved beyond its ambit to the suburbs. However, it is clear from the proposal that, following previous national exhibitions, Manchester's elite must have talked of the possibility of mounting something similar earlier than 1856. The men later involved in the Manchester Art- Treasures Exhibition would surely have been leading parties to such discussions and would have included members of the Botanical Society. Could the work in the garden in 1854–5 have been done in anticipation of such an exhibition?

It has been claimed that Gustav Waagen's *The Treasures of Art in Great Britain* published in 1854 inspired the Art-Treasures Committee – a copy was purchased for Manchester's Portico Library book stock on publication; the Portico loan book confirms that interested parties borrowed the book.[9] Dr Waagen had remarked that 'the art-treasures in the United Kingdom were of a character, in amount and interest, to surpass those on the continent' after he had travelled throughout Britain in 1851–2 visiting the owners of exceptional art collections. His visit to Manchester in the early 1850s was recorded in *Treasures of Art* as a failure. When he arrived he found his contact in the city was away, the weather was dreadful and, rather than wait, he immediately took the train to Liverpool where collections were described in detail. He did not return and Manchester is missing from his account. This no doubt explains the comment in the *Suggestions for an Exhibition*: 'Dr Waagen's valuable, but *imperfect* volumes'[10] (my italics). This slight to Manchester, and the long-standing rivalry with Liverpool, was surely a spur to the proposal to hold an Art-Treasures Exhibition in Manchester. It is impossible to say what occurred, but Society members were party to the discussions and could explain why the money spent in 1854 was exceptional in the context of the Society's finances. Minutes for 1856 and 1857 are missing and, given what unfolded, this may have been a political decision on the part of the Council.

Anticipating the thousands of expected visitors, the Council went to great lengths to implement improvements and the *Gardener's Chronicle* of 9 May included 'a few descriptive remarks' which indicated some of the work undertaken. 'There is a new walk leading direct [from the entrance of the garden] to the Art-Treasures Building ... twelve feet wide [4 m], in many parts overhung with trees, a pleasing vista ... [with] as termination the north transept'. A new fountain was installed in the centre of the Society's Exhibition House and a basin for goldfish on the adjacent terrace. On the lawn, 'where on exhibition days are assembled the aristocracy, beauty, and fashion of the district ... a monster tent 80 yards by 20 yards [73.2 × 18.3m], was erected for an American exhibition'.[11] The local press reiterated these preparations adding that military bands had been engaged and 'commodious refreshment tents' had been erected. Two extra tents had been needed because collections were being sent from the estates at Chatsworth, Derbyshire and Trentham, Staffordshire.[12] An addition to the lake was a pair of swans purchased at a local house sale – they bred so vigorously that they later had to be culled.

As some members were on both Committees they agreed there would be a physical connection between the two sites – a gate to the Garden.

As the Art-Treasures floor plan shows, 'On the North side are the Gardens of the Manchester Botanical Society and on the South the Altrincham Railway with a covered platform, 800ft [243.8 m] long communicating with the building by a corridor. The garden entrance, for pedestrians from the Exhibition site is clearly marked on the above plan, situated in 'the court on the north side of the transept'.

The finances of the Society were as precarious as ever. The accounts for January 1856 showed an overdraft of £807 18s 3d in addition to a £5000 mortgage and £1500 in loans, despite which the Council described the situation of the Society to subscribers as 'prosperous.'[13] The question that immediately arises is: where was the money to come from to fund these extravagant displays? As noted, the minutes for this crucial year are missing.

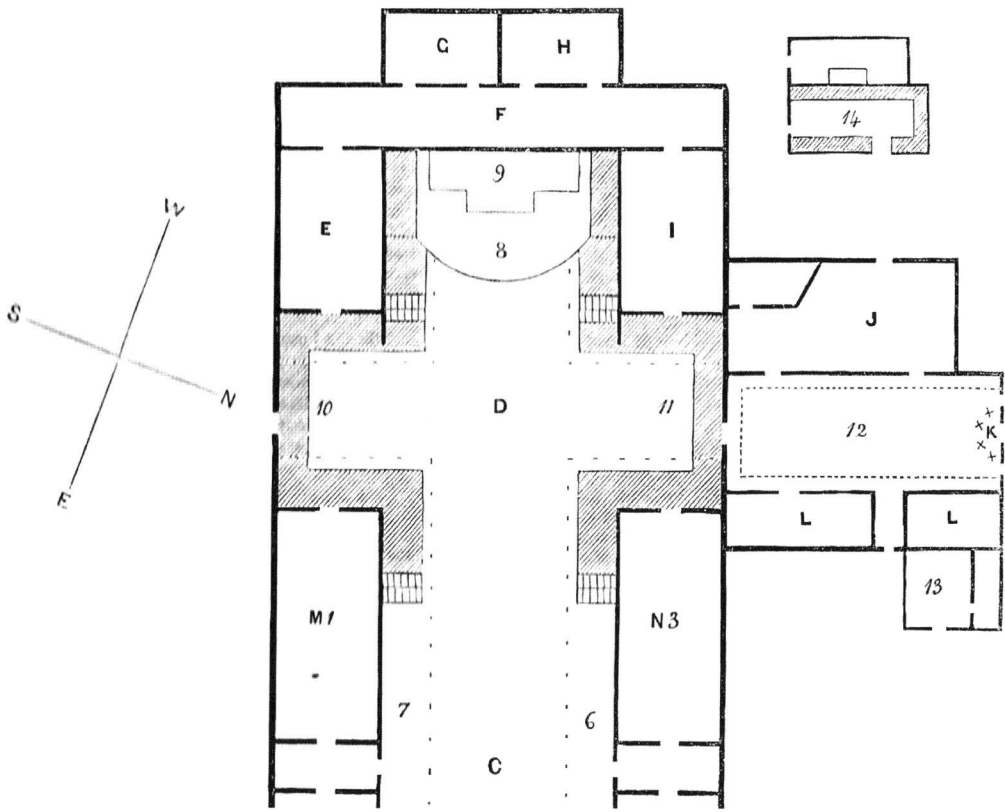

However a lavish brochure was published entitled *Manchester Botanical and Horticultural Society The Exhibitions of Plants, Flowers, Fruits, &c., for the present year.* This included a list of subscribing exhibitors, including the Queen, which suggests that finance was partly from voluntary subscriptions.[14] Three special exhibitions had been planned. On 26 and 27 May, it included the magnificent collection of American plants supplied by Messrs Waterer and Godfrey of Knap Hill Nursery, Surrey.

A second successful exhibition, which included the 'best collection of orchids ever seen outside the metropolis' as the stars of the show, was held on 24 and 25 June.[15] The third show, held on 26 and 27 August, was described as 'moderately successful.[16]

Although the dahlias were such a collection as 'never before seen at the Botanical Gardens', many entries were considered 'not in very fine condition'. A letter sent to members and possible competitors for the third show, from John Shaw, the Secretary to the Society, on 30 June demonstrates the growing financial difficulties:

> I am directed by the Council of the Manchester Botanical and Horticultural Society to inform you that in consequence of the great loss sustained by the Society by the two former Exhibitions, they regret to state that they are obliged to reduce the amount of Prizes to be given at the forthcoming Exhibition.[17]

In July the Treasurer, Mr Butterworth, had in fact recommended that this final show was cancelled on grounds of economy but the Council overruled him. An aside in the *Courier's* report of the first exhibition in May had already indicated serious problems:

> There was one drawback – no refreshment was to be had ... A curious scene might be witnessed at the gates near the Art-Treasures Exhibition, where refreshments were being handed through ... to the famishing and thirsty souls in the gardens, who looked like big birds being fed at the bars of their cages. What is the reason that Mr. Donald has abandoned his tents in the garden? We hope there has been no unpleasantness with the council.[18]

There had indeed been unpleasantness as 'certain gentlemen', who did not like the situation of the tents, had complained to Mr Donald, the caterer to the Botanic Garden Exhibitions. His reply was to the point, 'I have had nothing but dissatisfaction since I have had anything to do with you [the Council] ... I'll have them [the tents] away altogether.'[19] He did within two hours and the Council lost his premium of £100 for being awarded the catering contract – Donald did not suffer, he was caterer to the Art-Treasures Exhibition too.[20] This was merely an indication of the financial disaster about to face the Society. The Art-Treasures Committee instituted a rule change. Anyone leaving the Exhibition could not return without further payment. Financial disaster faced the Society. Shilling visitors would be deterred from visiting the Botanic Garden – the very visitors who would bring the profits. All the Society's efforts were thwarted by the 'no return' rule. As one member commented, 'Art triumphed over Nature, and the Botanical gardens lost their money.'[21] There had to be a reckoning. Someone had to take the blame.

After the debacle of the Art-Treasures Exhibition the inner workings of the Society were about to be exposed to the readers of the local press following a series of special general meetings. At the Town Hall on 23 September the first meeting was held to discuss the Society's financial problems and the press was out in force. They revealed a Society in disarray.

The Council explained they had expected that, with the draw of the Art-Treasures Exhibition, the money from the increased numbers of visitors to the Garden would allow them to liquidate the debts. To draw the public in, the Council had authorised a portion of the wall to be taken down next to the gate, replacing it with the railings, and exposing 'that bright prospect beyond': a strategy 'wooing only to betray'.[22] A member thought it would have been better 'if the dead wall had remained as it had done before' as they now had to pay for reinstatement. The Art-Treasures Committee's decision to institute the 'no-return without payment' policy is surely the most puzzling event in the whole affair. As the Botanical Society and the Art-Treasures Committee shared common members, why was the Botanical Society penalised in this way? It becomes clear on reading the local press reports that Butterworth, the Treasurer, was an extremely difficult man to deal with and that the Botanical Society Council was divided into warring factions over how to handle the

A flowering branchlet of *R. Taylori*. Belonging to the Vireya group of Rhododendrons originating in the Far East. This was a Veitch hybrid with clear pink blooms developed prior to 1877.
AUTHOR'S COLLECTION

Dahlia 'Dr Frampton'. From *The Floral World and Garden Guide A Complete Manual for the management of the Garden, Greenhouse, and* Conservatory (London 1858); Author's Collection. 'The Dahlia was first introduced into this country from Spain in 1789, by the Marchioness of Bute. This importation, and another made by Lady Holland, in 1804, were however lost to cultivation. A third stock was afterwards brought from France, about the year 1815, and from this the numerous forms have been obtained' G. Nicholson, *The Illustrated Dictionary of Gardening A practical and scientific Encyclopaedia of Horticulture for Gardeners and Botanists*, 431 (London 1884–8).
AUTHOR'S COLLECTION

debt. Returning, for example, to Mr Donald, the caterer, it is clear that Society members could be high-handed. Perhaps the continual harassment of the non-Society members of the Art-Treasures Committee by the high-handed demands of their Society colleagues, by demanding that their expenses in preparing the Garden should be re-imbursed, led to the withdrawal of their co-operation. No evidence is yet forthcoming.[23]

Other losses, too, were blamed on the Art-Treasures Committee. Even the Exhibition of American plants became a problem. The Art-Treasures Committee had enquired whether Waterers would mount a replica exhibition to their London show in one of the proposed side galleries. Hearing the news, the Council were appalled, a wonderful display of plants inside the Art-Treasures Exhibition would have a detrimental effect on visitor numbers to the Garden. Representatives rushed to London and engaged Waterers forthwith. The Council explained to members that they had understood that the Art-Treasures Committee would pay for all their expenses but their claims were rebuffed again and again. (As late as 1863 the MBH minutes of 15 November, reported that Butterworth had again been to see the residual Art-Treasures Committee to seek to recover the Society's costs.)

At the end of the meeting, suggestions were invited to ameliorate the debts, including selling the Garden, obtaining a further mortgage, and the previously failed device of raising money from the subscribers. Sir John Potter, MP hoped 'the public-spirited citizens of Manchester would … no longer allow the gardens to be aristocratic gardens … [and they] would become public gardens for Hulme.'[24] After heated exchanges, the members appointed a committee to examine 'the deplorable state of the Society' and report to yet another special meeting. This took place on 17 November, with 60 members present and three ways of liquidating the debt were suggested: an increase in the mortgage, an increase in subscriptions or the sale of the property.[25] The reported debate shows what attractions the gardens offered to shareholders; it was 'more thought of as a promenade', whilst for others the main attraction was the glasshouses. Exclusivity was still apparently the main benefit. A proposal that the Garden should be opened on Mondays and Saturdays to the public, thus raising an annual income of at least £400, was dismissed out of hand; the Garden was a private garden for the members. During the debate members of the Council were arguing amongst themselves and a Mr Romney claimed 'no-one had any confidence in the Council' and Mr Penney, agreeing, called for them to resign. A Council report concluded that, 'a too sanguine estimate of the benefits to accrue … from the proximity of the Gardens to that grand Exhibition' had been the cause of the problems.[26] A meeting on 2 December passed a resolution by 37 votes to three that the subscription rates should increase from 1 January 1858.[27] Events then took an unexpected turn when the Treasurer announced that he held guarantees of £1065 towards a fund to save the Garden and that shareholders should offer him protection when considering the contents of a letter published in the *Guardian* the previous Saturday, 27 November.[28] Michael

Potter, Chairman of the Society, had written the letter and he was defending himself against personal criticisms made at the special meetings.

This new dispute, generated by the letter, continued to entertain the local newspapers' readers. Potter revealed yet more expenses incurred by the Society in anticipation of substantial financial success. All the paths had been gravelled, five turnstiles for the shared entrance purchased at a cost of £200, a flag and flagstaff at 11 guineas, £90 was spent on policemen for the shows and the Society's premises had been moved from 'a dingy office in Princess Street' to 'aristocratic quarters in St. Peter's Square' where £33 had been spent on new furniture. In addition, a member of the Council was appointed managing director of the project for £150 for six months; Potter explaining that he had previously been offered the post for £250. The instigator of all this expenditure, Potter claimed, was none other than Mr Butterworth, the Treasurer, who had assured the Council at the outset that 'an influx of visitors would repay the outlay'. Potter felt that a man with a 'want both of judgement and temper' should go because 'as the author of the problems, how could he advise on the best mode to escape' the difficulties. When the final special meeting was held on 14 December, Butterworth was still Treasurer.[29] A report, signed by Potter, confirmed the rise in the annual subscription had been adopted and he claimed the Garden would be managed much more efficiently to become 'a credit and an ornament to the city of Manchester.'

Notes

1   W. E. Axon, *Annals of Manchester*, 271 (Manchester, 1886).

2   'Manchester Botanical Gardens The American Plants', *Courier*, 6 June 1857.

3   Letter, July 3, 1854, H.R.H. Prince Albert to Lord Ellesmere, *The Albion, British, Colonial, and Foreign Weekly Gazette*, N.S. 15(3), New York, August 2, 1856, 369.

4   Manchester Botanic Garden had gardenesque beds cut into the lawns. Joseph Paxton also described these geometric beds as the 'creations of art'. See: *Magazine of Botany* 11, 257 (London, 1844); D. Dewing (ed.), *Home and Garden Painting and Drawing of English Middle-class Urban Domestic Spaces 1675–1944*, (London, 2005); L. H. Albers, 'The perception of gardening as art', *Journal of the Garden History Society* 19, 163–74 (1991); R. Strong, *The Artist and the Garden* (New Haven, 2000).

5   J. Reilly, *The History of Manchester*, 469 (London, 1865). Membership has been confirmed using the membership lists, annual reports and entries in the minutes.

6   Reilly, *ibid.*, 470.

7   *Ibid.*, 471.

8   *Ibid.*, 469 and 472.

9   See G. F. Waagen, *The Treasures of Art in Great Britain*, 3 Vols., plus 1 supplementary (London, 1854); *ibid.*, vol. 3, 229.

10 *The Art-Treasures Examiner A Pictorial, Critical, and Historical Record of the Art-Treasures Exhibition, at Manchester in 1857* (Manchester, 1857).

11 Acid loving plants from North America, which included rhododendrons and azaleas, were extremely fashionable; many also came from the Himalayas and Far East.

12 *Courier*, 23 May 1857.

13 *Report of the Council of the Manchester Botanical and Horticultural Society 3 March, 1856* (Manchester, 1856) MBHS, MBH 3/2/20.

14 *Manchester Botanical and Horticultural Society The Exhibitions of Plants, Flowers, Fruits, &c., for the present year*, (Manchester: 1857), MBHS, 1638/1 (Local Studies Library, Manchester Central Library, Manchester). This brochure included a list of subscribers, which suggests that finance was partly from voluntary subscriptions.

15 'Manchester Botanical and Horticultural Society', *Courier*, 27 June 1857.

16 'Manchester Botanical and Horticultural Society', *Courier*, 29 August 1857.

17 Quoted in *Courier*, 29 August 1857.

18 *Courier*, 23 May 1857.

19 'Manchester Botanical and Horticultural Society', *Manchester Examiner and Times*, 24 September 1857, MBH 3/2/26.

20 See *Art-Treasures Examiner*, The Refreshment Rooms, p.v. The First Class Dining Room was 96ft long by 72ft wide (*c.* 29 × 22mo, a conservatory decorated in a Moorish style. The kitchen is described in detail

21 'Manchester Botanical and Horticultural Society', *Manchester Examiner and Times*, 24 September 1857, MBH 3/2/26.

22 *Courier*, 26 September 1857.

23 Some visitors did enter the Garden for other reasons; Tennyson was seen by Nathaniel Hawthorn having a smoke, as there was a no smoking rule at the Art-Treasures Exhibition. N. Hawthorne, *The English Notebooks 1856–1860*, 353 (Columbus, Ohio, 1997).

24 *Courier*, 26 September 1857.

25 'Manchester Botanical and Horticultural Society', *Manchester Examiner and Times*, 18 November 1857, MBH 3/2/27.

26 *Report of the Committee of Investigation into the Affairs of the Manchester Botanical and Horticultural Society, November 18 1857* (Manchester, 1857), MBHS, MBH 3/2/25.

27   'Manchester Botanical and Horticultural Society', *Manchester Examiner and Times,* 3
     December 1857, MBH 3/2/30.
28   Letter, 'To the shareholders of the Manchester Botanical Gardens', *Guardian,* 27
     November 1857, MBH 3/2/29.
29   'Botanical and Horticultural Society Special Meeting', *Manchester Examiner and Times,*
     15 December 1857, MBH 3/2/33.

CHAPTER TEN

'… The Darkness Before the Dawn'[1]
1858–1865

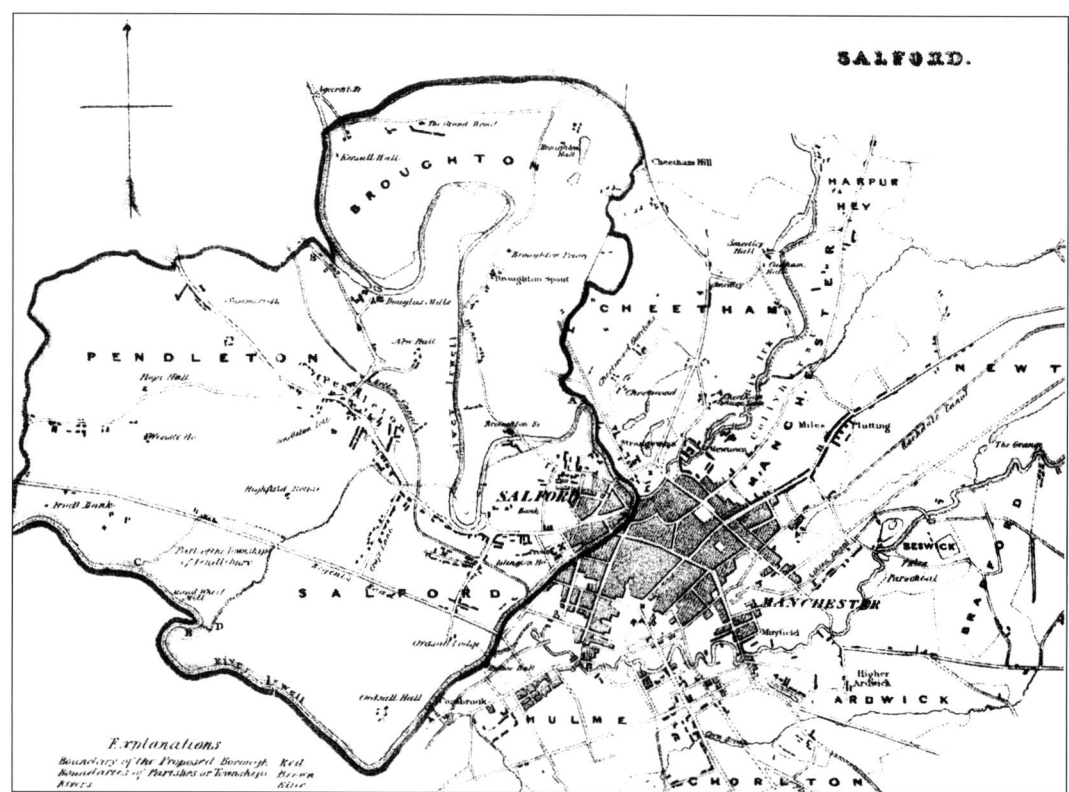

The year 1858 saw little improvement as other problems faced the Society and entered the public arena. Stephen Heelis, a member and the Society's solicitor, had commented at one of the special meeting that there was a 'kind of "cold-shoulder" feeling towards the gardens'. This suggests that, especially after the débâcle of the Art-Treasures Exhibition, the middle class were losing interest in supporting the Society.

By the 1850s the move of the Manchester upper middle class to the suburbs beyond the town's original boundary was also beginning to affect the Society. This was confirmed by Heelis who, as an inhabitant of Pendleton, wondered whether anyone living 4 or more miles [6.4 km] from the garden would be willing to pay the increase as 'they got very little in return', especially as 'there were plants in the houses a disgrace to the place'.

R. K. Dawson, Map of Salford, *Parliamentary Representation: reports from the Commissioners on the Proposed Division of Cities and Boundaries of Boroughs, Vol. II part I.* Ordered by the House of Commons to be printed 20/1/1832.

His comment on the condition of the plants brings us to the most notorious dispute in 1857. A report of the Garden sub-committee, published for members on 3 September 1857, had concluded that the Curator should be charged with 'gross ignorance and mismanagement' and that he should be replaced.[2] In the *Report,* the authors, Michael Potter, Francis Hernaham and Henry Micholls, lambasted the care of the plants in the garden. Their conclusion was that a charge of 'gross ignorance and mismanagement' should be brought against Campbell and a new Curator appointed.

> These means [of cultivation] do not appear to have been understood and practised by the Curator ... and therefore they are of the opinion that the time has arrived when a complete and thorough change should be made ... so as to render it more attractive to the public, and more beneficial to shareholders.

There had been murmurings against Campbell for some time. At the special meeting on 23 September 1857 it was noted that, 'there was strong feeling on the part of the anti-Treasurer party against Mr. Campbell, and foremost in opposition was the distinguished and immoral [*sic*] orator Mr. M. Potter, who had spoken so flatteringly of Mr. Campbell ... in the June'. Potter, in his letter in the *Guardian* on 27 November, had expressed surprise that Butterworth advocated Campbell's retention 'when it was a matter of notoriety that some two years ago, he was most anxious to remove him ... and make him Secretary, on the grounds of his inefficiency as curator'. *The Report on Plants and Gardens* makes clear the feelings of the authors: '... if any old indifferent variety of geranium is wanted, the most likely place to find it is in the Manchester Botanic Garden.'

Campbell must have been devastated by this report, as the minutes and extracts from published material have shown that he had received nothing but praise for all the work he had done over the years, including the work for the Art-Treasures Exhibition. His *Counter-Statement,* which was polite and considered, refuted the *Report* point by point: Amusing as this statement may seem (concerning indifferent geraniums), '... the trifling sum of a shilling has not been laid out on the purchase of a geranium since I have been connected with the garden ...'. He had worked as Curator for 25 years and had never bought a single plant, and claimed that £12 would cover all other expenses – certainly donations of plants to the Society had continued from its inception. Finally, he told the Council that a man who could read the *Report* and 'with hand on heart and declare before God, that it is *temperate, impartial, and gentlemanly,* must have a mind utterly at variance with my ideas of what is honourable and just'.

To Campbell's *Counter-Statements* was appended a report from Alexander Forsyth, whom Campbell had called as an independent witness. Forsyth had been gardener to Lord Stanley at Alderley Edge and the Earl of Shrewsbury at Alton Towers, and had seen the garden develop 'its beauties' over the previous 20 years and admired the way it was maintained by Campbell and his staff. He had read the *Report* and questioned the staff. He found that some of the

THE BOTANIC GARDENS.
To the EDITOR of the MANCHESTER COURIER.

Sir,—When these gardens were formed, it was by gentlemen in the neighbourhood of Manchester who had establishments, forcing houses, and collections of rare plants, from which the Botanic Gardens were, in a great measure, stocked; and I believe a large portion of the plants were presents from gentlemen amateurs, for the funds were always at a low ebb; and it is a practice with gardeners to exchange when they can. But there is one strange feature in the present management of the gardens, that there are only three gentlemen on the council who know anything of the cultivating of plants, viz., Mr. Hernaman, Mr. Nichols, and Mr. Potter. If this is not correct it is easily set right, but so it is currently reported. Will Mr. Barlow and Mr. Butterworth say what amount of knowledge they have of horticulture to entitle them to take a prominent position there?   AN AMATEUR.

Letter from 'An Amateur' to the *Manchester Courier* (26 September 1857)

ill-treated plants had been ordered to be pruned by members of the Council – at the wrong time of year – and were being nursed back to health. It was obvious from his experience that the sub-committee knew little of current advanced horticultural practices. If the Botanical Garden failed he was certain that Campbell was in no way responsible.

Letters appeared in the press commenting on the affairs of the Society. A pseudonymous correspondent, 'An Amateur' questioned if there were gentlemen on the Council who knew anything about gardening. A letter in reply from another anonymous writer, 'Truth', claimed to know all the men involved and said none were acquainted with plants. Their publication was miscalled '*A Report on Plants*' and he hoped the present Council would be removed from positions they were obviously unqualified to fill and, as a result, the Society would prosper.[3] To no avail! Campbell was awarded an annuity of £52 and appointed to take care of the Herbarium, re-named the Museum, and in this capacity his name continued to appear in the minutes. When the Museum was closed in 1862, Campbell finally left the service of the Society.[4] A new Council was appointed to regenerate the Society but it is clear from the politics involved that many old ideas were carried over into the new regime.

The re-election of some of the previous Council members did not engender confidence in subscribers that the Society would change. The poorly attended Annual General Meeting for 1858 considered the accounts for 1857.[5] The deficiency, excluding the mortgage, was £2885 4s 1d, an increase of £1581 16s 10d on the previous year. The Council decided that the annual exhibitions must continue, as they attracted new proprietors, but cheap admission would be discontinued except at Easter and Whit week. Presumably, it was felt that the return to exclusivity would help to attract new subscribers. An election for a new Council was the next item on the agenda and exhibiting gardeners proposed that Messrs John Shaw and R. S. Yates, both local nurserymen, should be co-opted as this would ensure men with practical knowledge were on the Council. The new Council would have 12 members though Butterworth, the previous Treasurer, argued for six on the grounds that 'he and his friends, if alone,' would have made better arrangements with the executive of the Art-Treasures Exhibition. At this point there was a rancorous and personal discussion between Butterworth and Potter, which the *Guardian's* correspondent declined to report. The new composition for the Council, passed by a two-thirds majority, was a President, three Vice-presidents and nine other members. The meeting degenerated into argument as the election proceeded. A Mr Barlow claimed that he represented many members in saying new blood was needed and that new, active and energetic men should be elected and it was unseemly that previous members were proposing themselves for re-election. Several subscribers then refused to stand if members of the old Council could be elected. Dr Ashton, who had agreed to be Treasurer, refused to serve with the second nine proposed, on the grounds they were commercial men who did not understand horticulture. The third suggested list included Shaw and Yates but was objected to on a technicality. Shaw then declined to serve and the election took place. Yates was not elected and five of those elected, including Butterworth, had been members of the previous Council though Michael Potter claimed he left 'without regret'. Although not in a majority, the re-election of some of the former office-holders did not augur well for a change of direction. The new Council published their intentions in the local press and claimed:

> ... after so many and such varied attempts to gratify the public taste in their management, [the Garden] cannot be made self-supporting without large and hazardous occasional outlay, their usefulness must be impaired, and their continuance rendered uncertain, if not undesirable.[6]

Membership was certainly declining and they wanted the new Curator, when appointed, to make the Garden an attractive daily resort and ensure science was promoted together with musical Promenades and Flower Shows. If shareholders refused to pay monies owing, the Council would advise the sale of the Garden. The Council had attributed many pecuniary problems to running exhibitions, and the author of a letter to the *Manchester Guardian* was appalled to see that there were to be three the following year, when the Council had promised only one. This behaviour was typical of the Society; exhibitors had complained of their treatment in the past and might not support the shows. He complained that some of the profligate remained, chaos could return and that it was open to conjecture whether the new Council could make a break with the past. These comments raise the possibility that the high-handed treatment shown by the Council towards Campbell may have been matched by a similarly cavalier attitude in the past towards the subscribers and exhibitors; if this attitude still prevailed, the author was indeed correct in his analysis.

By 3 May 1858 the new Curator, Bruce Findlay, was appointed to a Society riven by factions and deeply in debt. He found at once that the main concern of the Council was the outstanding dues of the subscribers: about 25% of the membership. The Council then approved the sale of annual tickets at a cost of one guinea. They were a contentious issue amongst Council members as

An Aside: Bruce Findlay, Curator 1858–1896

Bruce Findlay was born in Streatham, Surrey in 1835. His early training took place at the Tooting Nurseries of Rollinson and Sons, Surrey and from there he went to Kew to widen his experience. Next, Findlay moved to be foreman to the Botanic Garden in Hull and then to Sheffield, also as foreman, a position he held for two years. In May 1858, he became Curator of the Manchester Botanic Garden in Old Trafford. After sixteen years as Curator, Findlay was appointed to the dual role of Secretary and Curator and became the driving force behind the improvements in the Society's fortunes. His introduction of the National Horticultural Exhibitions enabled the Society to pay off the debts. His contributions were acknowledged by a grateful Society and in November 1881 Findlay was presented with a gold watch and chain and a cheque for £1000.

Findlay's motto was 'A garden is health, a garden is wealth, a garden is happiness'. The Manchester Botanic Garden was a living example of his belief and he was credited with providing such beautiful spectacles at the Garden that in the 12 years prior to 1881 at least 80,000 'shilling' visitors had paid to enter the

Mr Bruce Findlay. From *Journal of Horticulture and Cottage Gardening*, 1 September 1881, 206–7.
© ROYAL BOTANIC GARDEN, EDINBURGH

gardens.[1] In 1890 Findlay was presented with the Veitchian Silver Medal by the RHS. He was instrumental in founding the Manchester Horticultural Improvement Society in 1886 and became its first President.

The *Journal of Horticulture and Cottage Gardener* published a tribute to Findlay in 1881, which praised him not only as a gardener but also as a companion: 'Mr. Findlay is a gentleman of calm demeanour, sound judgment, extensive knowledge, great professional attainment, and never fails to accomplish what he undertakes … both rich and poor respect him. Amongst friends and acquaintances he is very social, and an excellent conversationalist'

Journal of Horticulture and Cottage Gardening, 1 September 1881, 206–7.

1   W. G. Baxter, 'Our Album Mr. Bruce Findlay', *Momus* 25 August 1881.

several thought their sale discouraged hereditary subscribers and so diminished the Society's income. Why pay two guineas if you could wait until May, and then pay a guinea for membership that allowed you free entrance to the Garden and merely prevented you from voting on Society matters? Indeed, some of the possible subscribers (either new or in default) may merely have paid the entrance fee for Exhibitions. The Council seems to have been unwilling to address the issue and by the end of May, 242 subscribers still had not paid. Butterworth had been missing from Council meetings since March; he was in Paris. In his absence, the Council debated whether to abolish the role of Treasurer, supposedly as it was a nominal position and required no knowledge of the state of the Society. This astounding remark ignores the Treasurer's responsibility to manage the finances though it is possible this approach was merely a device to facilitate removing Butterworth. By July Butterworth had still not been seen and finances were still being dealt with piecemeal. Butterworth was at last contacted and was unwilling to resign. A Memorandum was then inserted in the minutes:

> Arrangements for the last show – several members had strong personal feelings towards the Treasurer – as all parties want the shows to succeed it was conceived that the office of the Treasurer could be so arranged as on the one hand to avoid a compromise of Mr Butterworth's position and on the other to allay the irritations which appeared to exist.

Tensions prior to the election had obviously not been alleviated. Old enmities were still very much in evidence and focused, as before, on Butterworth whose personality and questionable competence was the heart of the problem. Though the Council considered appointing one of the banking firms in the city to the role, no action was taken. Certainly, they needed financial advice, as the minutes of 28 July 1858 recorded that the Lancashire Insurance Company were requesting immediate payment of their mortgage loan. Instead the Chairman and the Secretary agreed to ask for the mortgage to be raised from £6000 to £8500 on the guarantee of 'Gentlemen members of the Society' and the extra monies could then be used to pay the bank, Mr Buckley and the debts outstanding prior to 1858. The Lancashire Insurance Company however declined to extend the mortgage, the bank would lend up to £500 but would want guarantees and a member, Mr Harrison Blair, a Manchester barrister, had mentioned a client with £8000 to invest on good security (he later declined).[7] The Chairman, Secretary and Treasurer agreed to continue the search for an £8000 mortgage. There had also been a difficulty with the Trafford Estate. A letter of 28 July from Thomas Ayre, Sir Humphrey de Trafford's steward, reminded them that the erection of permanent refreshment tents in the garden was contrary to the provisions of their lease. To comply with Trafford's request, the refreshment tents were removed. It appears that the Trafford estate was constantly monitoring the Society's compliance with the terms of the lease and was still anxious to maintain the amenities of an area suitable for middle class housing.

With the financial problems still unresolved, the minutes of 18 September surprisingly record that the four men present agreed to major works in the gardens; a new nursery ground, trees felled, trees near the water removed, a new walk next to the Exhibition House (which needed repair), and an American ground or flower garden on the site of the maze. Yet by the beginning of October the financial situation was now a crisis. A letter from Hall and Jannison, solicitors for the Lancashire Insurance Company, threatened court proceedings if no instalment was paid on the mortgage and required that the interest owing be paid in full. By November, though the Council agreed to visit Hall and Jannison, they also asked Messrs Cowley and Cobbett, Manchester surveyors, to value the Society's grounds and property and the valuation were entered in the minutes:

> Valuation of the land containing 78,400 square yards [71,689 m²] at 2d per yard,
> £653 6s 8d *per annum.*
> Less chief rent to which the land is subject £120.   £533 *6s* 8d
> at twenty years purchase   £10,666 13s 4d.
>
> Valuation of the buildings consisting of Principal entrance and gateway, circular wing walls complete, Large Exhibition building, Conservatories, Greenhouses, Hothouses and pits and sheds, Heating Apparatus, Gardeners house, boundary wall. These I valued on the nearest calculation I can make of what they would realise if sold by auction or Private Contract to be removed from the site.   £3,000
>
> Total: £13,666 13s. 4d.[8]

The increase in the valuation of the land formed the basis for another attempt to secure a further mortgage. It was becoming clear that no loan would be forthcoming and other suggestions were made, including approaching wealthy townsmen, and the previously-failed device of a guarantee fund. Worse news was to come; the receipt of a notice from Hall and Jannison announcing they would foreclose and sell the Garden. The Council arranged to see the bankers, Messrs Loyd Entwistle and Co, and by the end of December, the Secretary submitted a bond of guarantee prepared for the bankers, which was signed by those subscribers present at the meeting (The Secretary was deputed to make sure the rest of the Council signed); the amount due to the Lancashire Insurance Company and other outstanding debts amounted to £1439 10s 5d By the AGM on 7 March 1859 the Council could announce that Henry Bury, Manager of Loyd Entwistle and Co, had granted them the loan. However Bury had accidentally seen the Council's report and regretted there was an inaccuracy. Though the report claimed that the loan had been secured for five years, it was not intended to be long-term and he did not want any future misunderstanding. Finally, the accounts were presented and showed that the Society had a surplus on the year of £716 5s 6d This optimism was due to the loan from the bank, though it was not mentioned that it was repayable on demand.

In an attempt to keep the finances in check, a financial report was presented to the Council on 19 May rather than waiting until the year-end. The income generated so far was stated as £995 2s though this figure included unpaid subscriptions of £382, but the liabilities were £8820 7s 5d This was a burden of debt that the Council could not support without some other sources of income. By the beginning of 1860 canvassing for new shareholders at last began in earnest. A circular describing the attractions of the Garden was to be left at neighbouring houses, to be followed by a canvasser. This applied to Old Trafford, Chester Road, Whalley Range and Stretford. The approach worked and new members began to be recruited. Findlay now appears regularly in the minutes, bringing more and more management suggestions to the Council. Findlay's first idea had been to approach owners of local businesses and allow them to sell tickets to their workers for Whit week. They were to cost 4d, not the usual 6d, but sold in bulk the sales should produce more profit.[9] Soon Findlay was preparing the schedules for the exhibitions, a job previously reserved to the Council. Next, he suggested two new exhibitions would help funds: a flower show in August with no prizes and a Chrysanthemum show at the end of November. The Council agreed to both.[10] 1861 saw further major expenses having to be dealt with: a new boiler for the Palm House and repairs to the Exhibition House. To subscribers, all may have seemed to be finally progressing satisfactorily. To the Council, Findlay's ideas must have seemed to hold the promise of a solution to the debts, which still loomed large.

On 8 January 1862 subscribers were given notice that the Treasurer was intending to propose the following resolution at the AGM on 15 January:

> That the following gentlemen be appointed a Committee to act with Messrs. Slater Heelis ... to consider the best mode of winding up the affairs of the Society. (Names to be submitted to the Meeting.)[11]

Butterworth also wrote a personal letter to all subscribers recommending the sale of the gardens.[12] In their annual report the Council admitted that yet again there had been a considerable loss of Proprietors. The Council wanted the final decision on the fate of the Garden to be taken by the shareholders many of whom, it was reported in the *Manchester Examiner and Times,* were in attendance.[13] The Chairman, Mr J. A. Turner, said 'he had celebrated the birth of the Botanical gardens; ... it seemed very likely they had now met to attend the funeral'. Some members proposed the gardens should be given to the Town Council as a gift to become an 'Albert Memorial' for Manchester. The Mayor of Manchester declined the offer there and then: not surprisingly, as it had been pointed out earlier that, with a debt of £9260, the City was unlikely to accept them. The Mayor had admitted to the meeting that 'it was a disgrace to Manchester that such Institutions did not flourish'. The *Guardian's* report also mentioned that many other institutions in Manchester, for example the Natural History Society, were also in financial trouble. This was the time of the 'Cotton famine' with trade severely depressed throughout the region.[14] One of the members, Mr Richard Haworth, commented that that 'even with bad trade surely they were not so dreadfully hard up as not to be able to raise £10 a share'.

Arthur Redford in *Manchester Merchants and Foreign Trade* argues succinctly that the periodic fluctuation in the cotton trade had a direct effect on life in Manchester. The relationship between the economy and the failure of public institutions had been suggested before. When Johann Georg Kohl, a German traveller, visited the city in 1844, he commented that:

> When I was in Manchester, most of the scientific, artistic, and literary institutions of the town (in which it was never wealthy), were in a very decaying state. The

MANCHESTER BOTANICAL GARDENS.

These Gardens are pleasantly situate, close to the Blind Asylum, at Old Trafford—a few minutes only by carriage drive or omnibus from the centre of the city.

They are sixteen acres in extent, and are most beautifully laid out, lawn and lake, forest trees and parterres of flowers alternating, so as to combine the grandeur of Park scenery with the delicate beauties of the richest Flower Garden.

The Greenhouses are very extensive, and are full of beautiful and rare Exotic Plants, affording to visitors the enjoyment of eternal summer.

At the back of the Greenhouses the different orders and species of Flowering Plants are placed in borders, and fully labelled, so that the student may, without difficulty, soon master the elements of the science of Botany.

In a Room over the Porter's Lodge is a Museum of Dried Plants, especially prepared to assist in the same study.

The Exhibition House has been erected at great expense, and is admirably fitted for the purpose of large competitive Exhibitions of Flowers and Fruits. Upon the whole, it may be safely asserted, that the Gardens are not excelled, if equalled, in Lancashire.

They are open to Shareholders and Subscribers every day except Sunday, from early morning till dusk, and the Officials will always be glad to furnish any information in their power.

By reason of deaths and removals, the number of Subscribers has, of late years, fallen below the paying point, and the Council, fully believing that the Gardens need only to be more fully known in order to be properly appreciated, respectfully invite you to visit them at your early convenience, on any, *except Exhibition days*, up to the 19th of April next.

The presentation of this Circular at the Lodge will procure your admittance with friends.

The price of the Annual Ticket is—

$$
\begin{array}{lllllll}
\textit{For One Person} & \ldots & \ldots & \ldots & \ldots & \ldots & \text{£1} \quad 1 \quad 0 \\
\textit{,, a whole Family} & \ldots & \ldots & \ldots & \ldots & \ldots & \text{£2} \quad 2 \quad 0 \\
\end{array}
$$

and persons who desire to be Shareholders may become so by payment of an additional Guinea for the first year.

The Gardens have been a source of great pleasure and also of great instruction for more than thirty years; they have done much to implant and foster the taste for Flowers which is now seen in the gardens of Gentlemen residing within a circuit of twenty miles from Manchester.

Zoological Society was selling its wild beasts by auction; the owners of the Royal Theatre had just been declared bankrupt; the Athenaeum was fast falling into ruins; … In a place where the utilitarian spirit of trade is so dominant, as in Manchester, such institutions are sure to be the first victims of any general depression in the commercial and manufacturing world.[15]

Was the crisis in the money markets in 1844 followed by the failure of the American cotton crops in 1845–46 the reasons for the Society's crisis in 1848?

Manchester Botanic Gardens Leaflet. Copy of the leaflet delivered by the Society to potential subscribers. Minutes, 14 March 1862, MBH 2/1/4.

The 'Cotton Famine' was clearly a contributory factor to the Society's problems in 1862 and appears to confirm Kohl's conclusion that social institutions were early victims of any trade depression.

William Slater, the Society's solicitor, confirmed that they could sell the land for building purposes though this would require the original covenants to be lifted. Councillor Rumney wanted to keep the gardens as he had no garden, and he agreed with a Mr Woolham that the gardens should be more exclusive not less, and the subscriptions consequently higher. Woolham pointed out that the Botanic Garden was 'the only private promenade in Manchester ... and worth 2 guineas a year'. A Mr Jackson agreed, 'Manchester was in many ways retrograding ... Our abominable architecture ... as well as the dispirited way in which they were cared for, was a disgrace to the community'. He was not surprised that many wanted to leave Manchester as soon as possible; another indication of the move to the suburbs. It was at this point that the suggestion was made that Garden could be saved by the introduction of Sunday opening. The special meeting duly reassembled on 21 February and voted on three proposals that, the Council should dispose of the Garden, the Garden should continue for another year at two guineas a family and the rules, that grown-up sons could not be admitted and that Sunday opening was not allowed, should be abolished.[16] At the Special Meeting on 26 February 1862, the Proprietors decided to continue the Society for another season at a cost of two guineas for families. If no extra support were forthcoming, and the Garden not to the City's credit, another meeting would be called in September.

The Council needed to raise money in earnest as they had promised the subscribers. New ideas were tried; nurserymen sold tickets for the Garden on a 5% commission, in the hope this would increase visitor numbers. Old habits still remained. When Professor Williamson of Owen's College asked whether his students could visit the Garden to study the plants he was to be charged £50 a year: Belfast College paid £50 a year to its Horticultural Society. Williamson replied that he would not entertain the idea.

The original founders had seen such study as being morally improving and a direct revelation of the divine. The new Council held other ideas. It did not re-introduce scientific beds as promised and continued to run the Garden as an exclusive pleasure ground (some botanical beds were installed behind the greenhouses). The Council did at last set about collecting subscriptions with vigour. Three collectors were appointed, their commission was 10% and within a week they had collected £240 9s.[17] Findlay proposed that he should collect the subscriptions himself and a proportion of the money raised would take the place of his advance on salary.[18]

It becomes clear that he felt his abilities should be utilised to save the Garden and he began advising the Council to this end. His reports increasingly influenced the Council's decisions as every month passed. He was taking a prominent management role, not merely in the organisation of the Garden but indirectly in the running of the Society. For example, Findlay on 11

An Aside: Sunday Opening

When the Manchester Garden was founded in 1831 there was no Sunday admission, as might be expected from a Society that had put religious beliefs as a cornerstone of its foundation. In Manchester the rule prohibiting Sunday opening first came under challenge in 1848, and the clamour for Sunday opening increased as the Garden came to be regarded as a pleasure ground rather than a scientific garden. In 1849, a motion calling for Sunday opening had obtained only one vote in favour. The campaign began in earnest at the Annual General Meeting on 3 March 1856; the motion proposed was 'That the gardens of the Society be open to the Proprietors on Sundays from 2.30 pm until dusk' and the arguments raised were several.[1] Those for the proposal argued that many members had no gardens and the Botanic Garden was perceived as their own in which they should be allowed to walk at all times. Some members worked six days a week and Sunday was the only day available to visit the Garden. Staff already worked in the Manchester garden on Sunday and, as this did not interfere with their churchgoing, any opening could be covered by the current arrangements.

The prime concern of those opposing the motion was linked to ideas known both locally and nationally as 'The Sunday Question'. Opening was seen as the 'start of a slippery slope' and it was argued that 'the public recognition of Sunday recreation and diversion, the sport, the theatre, the open shop, and the gambling house' would also be sanctioned. Lurid descriptions of degenerate behaviour on the streets of Manchester on Sundays only added to the argument that exclusivity was essential for the middle class. The opposition claimed that the actions of members would 'give a sort of official sanction ... to acts of Sabbath desecration' and in assuming a position of class leadership they felt the need to set an example. If they sanctioned opening the Garden then there would be a demand for the opening of 'museums and art galleries'. The opposition went so far as to say that they would hear arguments in the House of Commons that 'Manchester is becoming strongly influenced in favour of Sabbath Desecration'. The members clearly saw themselves as a very select group who could influence events locally and nationally. At the close of the poll 655 votes had been registered and the motion was defeated by 69 votes. The Garden stayed closed

Guardian, 4 March 1856.

1   From a report in the *Manchester Guardian*, 4 March 1856.

August has procured the offer of a 10 guinea prize for orchids donated by B. Williams of London, and the Council accepted; Findlay suggested that, as the Manchester Athletic Club was meeting at Old Trafford on the same day as the show, allowing them admission at 6d would produce a large receipt. This idea was declined, exclusivity was still more important than a guaranteed source of income. Findlay's financial acumen showed an understanding of the underlying problems, but he seems to have felt it was inappropriate to challenge the Council directly on financial matters. The minutes of 28 September contain the accounts for six Promenades held during the summer; they had produced an income of £89 16s with only £12 11s profit. However, if the costs of the Exhibitions were added into the accounts, the profit was only £2 14s 8d. This must have worried him and he sought alternative ideas to solve the problems.

The Society's financial affairs had become a national scandal. William Robinson, one of the best known gardeners of the day, visited the Manchester Botanic Garden in September 1864 and interestingly, before describing the Gardens, Robinson describes the debt:

> The annual income is ... about £2,000, but £450 of this are annually required for interest of debt. So rich a city as Manchester should be able to support a fine botanic garden, and I have no doubt that if the incubus of debt were removed, the garden would speedily improve and be a credit to the city ... During the last seven years, the garden has been brought from a state of poverty of interest ... by the present curator, Mr. Bruce Findlay ... However, his efforts cannot liquidate a heavy debt.[19]

Further financial problems appeared in April 1865 when Mr Stern, Manager of District Bank, sent a letter to the Council:

> Be so good as to bring before the Board Meeting of the Botanic Society the balance due to this bank (about £3,558) ... it has been allowed to lie dormant for nearly two years, be so good as to send me a copy of your intent in this matter.[20]

Though the Council could not know it at the time, the dawn of a real recovery was about to begin. Under Findlay's influence Manchester was to become the 'metropolis of horticulture outside London'.[21]

Notes

1   W. A. Shaw, Botanical Gardens, *Manchester Old and New* 3, 50 (Manchester, 1894).

2   *Reports of the subcommittees appointed to inquire into the condition of the plants and gardens; together with the salaries and wages paid by the Society and the Counter-Statements submitted to the Council by the Curator, September 3rd, 1857* (Manchester, 1857), MBH 3/2/4.

3   Letter, *Courier,* 3 October 1857.

4   Minutes, 5 March 1862, MBH 2/4/1.

5   'Manchester Botanical and Horticultural Society Annual Meeting', *Guardian,* 2 March 1858, MBH 3/2/37. The report states few members were present.

6   *Address to the Shareholders,* Manchester, 8 March 1858 (Manchester, 1858), MBH 3/2/34. See also 'The Manchester Botanical and Horticultural Society', *Guardian* 10 March 1858, MBH 3/2/38.

7   Minutes, 18 August 1858, MBH 2/1/4.

8   Minutes, 1 November 1858, MBH 2/1/4.

9   Minutes, 3 May 1860, MBH 2/1/4.

10   Minutes, 6 August 1860, MBH 2/1/4. See also Minutes, 7 February 1861, MBH 2/1/4.

11   Letter, Notice of winding up petition, dated 8 January 1862, MBH 3/2/42. See also Letter, Resolutions to be put at the adjourned annual meeting 21 February 1862, MBH 2/1/4. Slater Heelis were the Society's solicitors; see *Slater Heelis and Co. bicentenary 1773–1973* (Manchester, 1973).

12   Letter, Butterworth, 'To the members of the Manchester Botanical and Horticultural Society', 18 February 1862, MBH/3/2/48.

13   'Manchester Botanical and Horticultural Society', *Manchester Examiner and Times,* 16 January 1862

14   For discussions of the 'Cotton Famine' see J. Watts, *The Facts of the Cotton Famine* (Manchester, 1866); W. O. Henderson, *The Lancashire Cotton Famine 1861–1865* (Manchester, 1934); A. W. Silver, *Manchester Men and Indian Cotton 1847–1872* (Manchester, 1966); J. Walton, *Lancashire: a social history, 1558–1939* (Manchester, 1987); A. Kidd, *Manchester* (Keele, 1993).

15   J.G. Kohl, *Ireland, Scotland and England* (London, 1844); see also A. Redford, *Manchester Merchants and Foreign Trade 1794–1858* (Manchester, 1934, reprint 1973).

16   See 'Manchester Botanical and Horticultural Society', Letter to subscribers, 14 February 1862, MBH 3/2/47.

17   Minutes, 5 March 1863, MBH 2/4/1. A circular was delivered prior to the visit describing the garden and its facilities, which makes clear it was closed on Sundays.

18   Minutes, 2 February 1863, MBH 2/4/1. This was 2% on the first £1000 of old subscriptions, 5% on the remainder and 10% on new subscriptions.

19   'Notes on Gardens, Manchester Botanic Garden', *Gardener's Chronicle,* 24 September 1864, 915–6.

20   Loyd Entwistle and Co. had amalgamated with the Manchester and Liverpool District Bank.

21   W. A. Shaw, *Manchester Old and New* 3, 50 (Manchester, 1894).

CHAPTER ELEVEN

'The Genius Behind the Machine'[1]
1865–1887

The Fairy Fountain in the Botanic Garden, Queen Victoria's Jubilee Exhibition. From H. Garside, *The Royal Jubilee Exhibition, Manchester 1887: a photographic record* (Manchester, 1877).
CHETHAM'S LIBRARY, MANCHESTER

On his own initiative Findlay began actively seeking ways to reduce the debt. In September 1865 he suggested a spring show might encourage new subscriptions, because attention would be drawn to the Society six weeks earlier than usual, arguing that similar shows had been held in London and Liverpool with great success.[2] With the prospect of the loan being recalled, the Council willingly agreed. They also took the opportunity in February 1866 to join an organisation proposed by the Royal Horticultural Society (RHS); a Union, which would 'offer advantages to Provincial Horticultural and Floral Societies' with an annual subscription of two guineas. The first International Horticultural Exhibition took place in London in 1866, organised by the RHS and representatives of government departments and botanic gardens. Although a member of the RHS Union, there is no indication that the Manchester Society was invited to take part.[3] However, this exhibition was to have a major impact on the fortunes of the Society; Bruce Findlay's visit produced a new idea. A National Flower Show should replace the traditional Whit Week Exhibition and this would create more income for the Society. The London Exhibition lasted four days and was visited by 82,000 people. Profits were over £5000, surely a point not lost on Findlay.

The Council demurred but this time he persisted and his letter of 6 September attached a list of Guarantors to a fund that would protect the Society against loss. A letter was also received from the *Gardener's Chronicle* to say that if the show was a failure, they would refund half the advertising costs. The Council finally agreed. Findlay's strategy was to increase the value of the prizes and by 7 December he had £980 promised in prize money; most prizes offered were between £150 and £250. This was intended to attract a greater number

An Aside: Exhibitions

Exhibitions of plants and flowers were not new. Florist and Floral Societies held them from the eighteenth century and they became an important feature of the gardening societies and subscription botanic gardens in the nineteenth century. The concept of national exhibitions took longer to develop. In 1844, national exhibitions were held in both France and Germany; showing exhibits which related to many aspects of the nation's life, both economic and cultural. This inspired the Society of the Arts to propose that a similar national exhibition should be mounted in Britain.[1] This was held in London in 1847, under the patronage of Prince Albert, as an exhibition of 'Works of Industry', and became the inspiration for the Great Exhibition at the Crystal Palace in 1851.[2] The Crystal Palace Exhibition

Manchester 1851. 'Gone to the Exhibition'. From H. Thornber, *The Later Work of George Cruickshank* (London, 1888).
B. HAWORTH'S COLLECTION

opened on 1 May 1851 and, though visited first by the wealthy of both Britain and the Continent, was soon to be open to the 'shilling' admission ticket with the crowds coming in their thousands by train from home and abroad. The Exhibition in Hyde Park was a sensation, from the glass palace designed by Paxton to the exhibits of science, art and industry. In 1861 the RHS opened its garden in Kensington and in 1862 their Great Spring and Summer shows began, together with specialist shows, such as the rose show and the tulip show. The Manchester shows, introduced by Findlay at the botanic garden, were examples of a national trend.

1 R. Drayton, *Nature's Government: science, imperial Britain and the 'improvement' of the world*, 193 (New Haven and London, 2000)
2 S. Gunn, *The Public Culture of the Victorian Middle Class*, 76–8 (Manchester, 2000)

of exhibitors from outside Manchester. A letter in the *Gardener's Chronicle* apparently confirmed this, the author claiming he had never seen such generous prizes and that they would ensure an enormous number of competitors.[4]

Exhibitions had been a part of the life of the Manchester Society from its establishment in 1827. At an exhibition of fruit and vegetables in October 1830, the prizes included the London Silver Medal for first prize and, from the Society, two silver cups for second and third prizes; as a corresponding member of the London Horticultural Society (later the RHS), they received a Banksian Silver medal to be awarded annually.[5] Manchester men were already showing their produce at the London shows; exhibitors at the Society's Whit Week Exhibition in June 1831, for example, were offered assistance in sending their entries to the London Horticultural Society's annual show the following week.[6] The shows were small, there were two judges and entries limited to the local area.[7] In 1839 they were still perceived in the context of science and religious belief as 'wonderful display of wisdom, power, and the goodness of the Being that created [them]'.[8] The Society developed a standard procedure over the years. Arrangements were made and judges appointed at the beginning of the year.[9] The event was advertised in the Manchester press and the *Gardener's Chronicle* for the preceding two weeks. A routine letter was sent to the bandmaster stationed at Hulme Barracks requesting the regimental band play at both Exhibitions and Promenades. This was not unique to Manchester. Hiring the local regiment's band was a common feature of subscription botanic gardens; for example, in 1816 the Liverpool Society, struggling to attract new members, advertised an evening performance by the band of the 54th regiment.[10] Prize money was another matter for discussion, although not all the prizes were paid from the Society's funds, some being donated specifically into a special fund; occasionally a member might donate a prize, for example, for one class of plants. Another feature of the later exhibitions became the inclusion of other societies' shows; for example the Auricula Society, the Carnation and Picotee Society and the Tulip Society. The Whit Week Exhibition was the main fund-raising event of the year and, although the entry price for the first day guaranteed exclusivity and attracted the wealthy patrons, the Council again recognised that many 'shilling' visitors' were also needed.

With this experience behind them, in 1866 Findlay and the Society began to organise the National Exhibition and 6 October 1866 the *Gardener's Chronicle* publicised the Manchester event to the nations' gardeners:

> The Manchester National Horticultural Exhibition is ... to open on Friday 7 June, 1867 and to close on Saturday the 15th. This period includes the Whitsun week; and being a holiday season throughout the manufacturing districts, the population of which are great pleasure seekers, there seems to be every prospect that the show may be financially successful ... a substantial guarantee fund has already been subscribed.

A letter to the *Gardener's Chronicle* published in November noted there were no prizes for roses, and supported the idea of 'shilling days' to draw in the crowds. The author concluded, no doubt to the delight of the Society's who hoped to make similar profits:

> ... it reflects much credit on the taste and liberality of the floral brotherhood at Manchester. *Quod facio, valde facio*, was the motto of the great commercial city when she collected together the Treasures of Art, and there is no declension in her munificence, now that she essays to assemble the masterpieces of floral skill.

The Council decided to pre-empt the weather and erect a large canvas tent, three times the size of the Exhibition House, so the whole event would be under cover. Perhaps an article in the same *Gardener's Chronicle* had influenced them as this had advised the Council to take note of the 'monster' show at Nottingham. The

Exhibition there had lacked any striking highlights as the use of multiple tents allowed only limited displays. A gardener employed at the Manchester Garden, Mr J. Forsyth Johnson, described the tent in *Gardener's Chronicle* in April 1867. Designed by Findlay, it was to cover 36,280 square feet [11058 m²] of which 19,080 square feet [5816 m²] were grass banks. The tent proved a triumph:

> The principal limb of this exterior space presented a very picturesque effect from a raised mound which commanded a birds-eye view of the whole. The parallelogram was laid out in a series of circles, forming concentric terraced platforms, and was principally occupied by Azaleas and Pelargoniums, while the sides form a series of grassy slopes [for plants].[11]

To maintain exclusivity, entry was to cost 10s 6d on the first day, 2s 6d on the second and 1s for the remainder. The published schedule indicated the prizes were generous.[12] Noblemen and national nurserymen would be exhibiting. Findlay informed the Council in May that they included the Dukes of Hamilton and Buccleuch together with several well-known nurseries: B. L. Williams, Holloway; Paul and Sons, Cheshunt; Ivery and Sons, Dorking; John Waterer, Bagshot; and Lane and Sons, Berkhampstead. It is clear from the minutes that Findlay was in full control of the arrangements, from hiring the bands to organising the refreshments. He also organised transport for plants to and from the show with the London and North West Railway Co at the price of a single fare; this was something the Council had attempted, and failed, to achieve in the past. Findlay submitted a list of 18 Judges to the Council, who included some of the most famous men in the gardening world; those from botanic gardens included: Thomas Moore, Chelsea; John Smith, Kew; John Ewing, Sheffield; and J. C. Niven, Hull.

There was one small problem; the dates of the Manchester and the RHS shows would clash. Mr A. Pettigrew of Brighton Grove, Manchester wrote to the *Gardener's Chronicle* about the attitude of the RHS:

> ... two great flower shows are fixed to come off at the same time. For many reasons this is to be regretted. Manchester folk have been ... making a noble effort to secure a grand show in June next. ... Both the Royal Botanic Society and the Committee of the York Floral Fete have evinced a friendly spirit in avoiding a clash with the Manchester Show.[13]

Findlay went to London to negotiate but there was no alteration. Fortunately, this did not deter the famous plant nurseries in London and the south of England from exhibiting their plants. A revealing explanation was given the following December when it became clear that the two shows would clash again. The RHS Great Spring Show was part of the London Season and the dates were constrained as 'Epsom and Ascot, and other metropolitan shows, left the Society no alternative'.[14] This point was reiterated by an article in a Scottish magazine, *The Gardener,* where the editor explained that the Royal Caledonian Society would hold an Autumn Exhibition of fruit, a date earlier in the year being unsuitable because:

The great London Societies are by force of circumstances compelled to have their exhibitions during the London season ... The English provincial Societies generally hold their exhibitions to suit the early summer holidays'.[15]

The Whit Week Show was part of the Manchester Season dictated by the Wakes-week holidays allowing the maximum participation by all social classes; unfortunately Manchester's early summer holiday clashed with the RHS spring show.

A June editorial in the *Gardener's Chronicle* shows the Society's finances were still of national interest as the editor expressed the hope that the National Exhibition had ameliorated 'the discredit and clog of a heavy debt'. Their July report assured readers it must have been commercially successful and the editorial compared the London International and the Manchester National exhibitions, concluding Manchester's display of Orchids was 'equal to the International, if not superior'. Findlay was credited for arranging such an enormous show in the provinces. The exceptionally detailed accounts of the event showed a balance in favour of the Society of £398 4s 3d (a considerably smaller amount than in London) though 50,000 visitors had seen the show. Focus must fall on the free entry of members, their families and friends as the reason for the difference in profit. Could the Society really afford this? In gratitude for the success, the Council raised Findlay's salary, backdated to the previous March, and organised a public subscription.[16] The *Gardener's Chronicle* announced the following month:

The Manchester Botanical and Horticultural Society has decided to hold another GREAT EXHIBITION next year ... nearly the whole of the Guarantors having consented to renew their guarantee for another year. It is intended to offer about £1,000 in prizes so that there is little doubt of a great show being got together; as the "Whit-week" is the only time when the Society can hope to "pull through" successfully, it is very desirable for all parties that there should be no clashing with other shows as happened last year.

In January 1868 Findlay was appointed to the uniquely dual role of Secretary and Curator. Under Findlay's direction, the Society was to enter its most successful years.

Arguments between the Manchester Society and the RHS seemed to have been overcome when the Society was informed in May 1868 that the RHS was to stage their next provincial Show in Manchester in July the following year. It would be a joint exhibition with the Society in the Botanic Garden. However this did not happen and the affair highlights the growth of horticultural exhibitions nationwide, the Botanical Society's perception of its own role within Manchester and the financial consequences of free entry for members. To extend the influence of national exhibitions, Prince Albert had suggested in 1861 that the RHS should stage provincial flower shows. By 1868 the RHS had staged two (Bury St Edmunds in 1867 and Leicester in 1868), both of which generated substantial profits. In June 1868 they sought shared funding from the Manchester Botanical Society, in return for which there would be a joint division of profits. The Council showed the letter to the Manchester branch of the Royal Agricultural Society and asked them whether they were willing to donate £600 to the venture in return for a share of the profits. They also wrote to the RHS to enquire whether subscribers would receive free admission. Discussions took place the following January to settle the prize fund but were not a success and the Council only donated £25. This was not dictated by financial problems, for in February 1869 the Society had paid £1785 to the bank to reduce the debt. In March a letter was sent to the RHS making clear that, though the Society was willing to guarantee the required £600, they would not be responsible for the special prizes. The Council, however, still required an equal share of the profits and on these terms they would place the Garden at the disposal of the RHS with one condition: the Society's members must be admitted free as they could not be charged entry to their own garden.[17] The RHS declined, pointing out that profits would be doubtful if members were free, so the talks failed.

The disputes between the two Societies drew the attention of the press, especially as the RHS then joined with the Agricultural Society's Annual Show to be held at Old Trafford. Findlay, in a letter to the *Gardener's Chronicle*, argued that Manchester subscribers had always had free entry to the Manchester shows and the receipts always showed a profit. He also reminded the RHS that great horticultural shows were not a novelty in Manchester and they had expected too much assistance from a Society whose own National Show was to be held in August. The RHS show was a disaster:

> The show itself, as a whole, may be said to have been one of average quality ... the local arrangements were anything but good. ... The agriculturalists were to have it all their own way. ... Poor horticulture was pushed on one side in a remote corner and access was on foot. ... it is still a fact that the arrangements were bad.[18]

When the Prince and Princess of Wales had visited on the Wednesday, they had been the main attraction, so much so, that any visitor who wanted to see the floral displays could do so 'with tolerable facility'; though there were many gaps as several exhibitors had not arrived. The failure was compounded for many by the expectation that the show would be a notable one due to the reputation of past Manchester exhibitions. Perhaps the RHS were naïve and patronising to choose Manchester in the first place. The Manchester Whit Week Show of 1869 was another resounding success and the report in *The Gardener* claimed; 'This is now *the* great Horticultural Exhibition of the Midland districts ... a gathering from all parts of the kingdom'. The Council seemed to have right on their side as the accounts were in profit and the debt had been reduced by a further £1000. In addition, as the income for the year was over £4000, the Society built a new orchid house at a cost of £230 14s 8d.

The two successful RHS shows in the provinces must have demonstrated the scale of profits they could expect if members and their families paid even a nominal fee. Exclusivity is the key to why this did not happen: subscribers had to have free access to their private garden. Though annual members had no such claim, to argue that they should pay an entry fee would have prejudiced retention and recruitment of such members. Unfortunately the Council did not seem to realise that they were dealing with changing times. Initially, the membership fees paid for access to an elite fashionable garden where small annual exhibitions were held. Previous causes of financial problems for the Society were solely related to running and improving the Garden for the membership. However, the growth of National Exhibitions changed this. The Manchester Society was now staging large prestigious exhibitions and this meant increased costs and increased visitor numbers, with the potential for greater losses. In effect the Garden became a stage where the importance of Manchester's gardening scene could be displayed and in that sense it ceased to be the members' private garden for the duration of the Exhibition. The need to maximise profits was essential both to cover the exhibition costs and as a hedge against financial problems. It was true, as Findlay said, that the Manchester exhibitions had recently shown a profit but the minutes show that these were often minimal. The Council seemed to have neglected, or refused, to acknowledge that new considerations applied to staging events on a national scale. There was perhaps little difference between the exclusivity of the RHS whose members, headed by the aristocracy, were concerned with the London Season and a desire to extend their patronage to the provinces, and the attitudes of the middle class members of the Manchester Botanical Society who had links to the local nobility and who were determined to keep the garden as an exclusive club whatever the cost. Their patronage extended to the working class, allowed access under specified conditions. Continuing to allow members free admission to the exhibitions perpetuated this myth.

A description of the Manchester Garden appeared in the *Journal of Horticulture and Cottage Gardener* in 1871. J. Robson, the author, agreed with William Robinson that by this date the term 'botanic garden' dignified a fashionable resort where showy plants were grown in the summer months. Few places still grew plants of a permanent botanical nature though he credited

By the lake in the Botanic Garden. From H. Garside, *The Royal Jubilee Exhibition, Manchester 1887: a photographic record* (Manchester, 1877)

The Manchester Botanic Garden 1871

The appearance of the imposing-looking gate and lodge, which strikes the passerby as being a likely entrance to some place of importance, is further enhanced by another gate (differing in detail but yet artistic) on the other side of the road, leading to the mansion of Sir Humphrey de Trafford, Bart. Unfortunately the presence of houses on both sides destroys the quiet seclusion of the garden, and their rapid approach threatens at no distant day to swallow it up. Even now it is hemmed-in nearly on all sides, and the difficulty of growing good specimens of shrubs and trees is increasing every year … we must remark that everything is being done to mitigate an evil which no one deplores more than the worthy curator, Mr. Findlay. Indeed, the wonder is how he manages to present such a healthy collection of exotic plants under glass, and also to furnish such a rich display in the flower beds, which are both numerous and full-sized.

… On entering the garden, the first thing we see is a large and somewhat open grass plot. On this the Manchester horticultural shows are held. There is a peculiarly-constructed glass house of large size, some of its details certainly having an oriental rather than an English outline. Adjoining and leading out of it, the ground has been formed into a series of turf covered stages for plants, much in the same way as at the London Botanic Society, Regent's Park. Even at the time of my visit Mr. Findlay had turned some of the circular stages to account by planting upon them Gladiolus and other showy plants, which were doing well, and showed to advantage. Fronting this was the flower garden proper – a series of beds laid out on grass. This geometrical garden was, of course, planted with ordinary bedding plants, which by conventional usage are supposed to be best adapted for such purposes: and other beds – scattered promiscuously over the ground, contained plants interesting to the lover of novelties as well as the studious botanist. The ground in this direction is bounded by a piece of ornamental water and some shrubbery, with some rustic-work, which only requires the shrubs to thrive better in order to look well.

Pursuing our walk to the left we come upon the range of plant houses. Plant stoves are also to be seen, and two or three Orchid houses lower than those we have been in before. … We next follow him into some Fern houses. The whole collection, in-door and out, was in a most praiseworthy condition for such a place, and if the shrubs were not all that could be desired, it was cause beyond control. Mr. Findlay may justly claim a small amount of credit for the good condition of the plants in the houses, and also for the flowerbeds on the lawn.

J. Robson, *Journal of Horticulture and Cottage Gardening*, 17 December 1871 (abridged)
© ROYAL BOTANIC GARDEN, EDINBURGH

Manchester as one such example.[19] His article draws attention to the points already raised as possible problems for the Society: the pollution caused by a large industrial town; the growth of housing around the garden; the move to the suburbs where there was a better atmosphere; and lack of interest in scientific planting.[20] The same year saw Alexander Forsyth write:

> The well-to-do portion of society seek for high-class amusement, and will come a long way to promenade in the seclusion of the Botanic Garden, where the half-guinea entrance on the opening day ensures the select character of the visitors.[21]

NATIONAL HORTICULTURAL EXHIBITION.

1874.

This Exhibition continues to be the most important feature in the Society's transactions, the one held during the past year being equal in point of merit to any of its predecessors. It will be seen, by the following list of Exhibitors, that the most important growers in the country compete for the prizes offered in the schedule; and by referring to the balance-sheet, it will be seen that these Exhibitions are highly appreciated by the public, and it is hoped that, by making them more attractive, to secure a larger measure of public support. The importance of these Exhibitions must be obvious, inasmuch as they tend to the improvement of Horticulture by rewarding superior excellence in practical gardeners; by exciting emulation among them; by introducing, through them, new Plants, Fruits, and Vegetables, or by increasing the cultivation of such as had fallen into unmerited neglect. The weather was upon the whole favourable, and therefore the attendance of visitors good. The sum of £789 was awarded in prizes. This may seem a large amount; but when it is remembered that the Exhibition extends over seven days, it must be apparent that this sum is comparatively very small. The Exotic Orchids again formed perhaps the most striking feature in the Show. The magnificent specimens of Roses in pots excited the admiration and astonishment of all who saw them. The miscellaneous Stove and Greenhouse Plants were marvels of culture; whilst the stately Palms, graceful Ferns, ornamental Dracænas, gorgeous Rhododendrons, curious Japanese Plants and Yuccas, and interesting Alpine Plants, combined to form a scene which it is hoped afforded pleasure and instruction to all visitors.

Among the Amateurs who took prizes in the classes in which they competed, were:—For Orchids, Dr. R. F. Ainsworth, of Lower Broughton; Messrs. O. O. Wrigley, of Bury: E. Wrigley, of Bury; Joseph Broome, of Didsbury; and W. Hadwen, of Fairfield. For Stove and Greenhouse Plants, H. Samson, of Bowdon: John Rylands, of Longford; John Stevenson, of Timperley; and John Potts, of Whalley Range. For Stove and Greenhouse Ferns, O. O. Wrigley; T. M. Shuttleworth, of Preston; and J. B. Mason, of Cheetham Hill. For Hardy Ferns, Miss Pearson, of Prestwich; Henry Crowe, of Greenheys; and T. M. Shuttleworth. For Dracænas, Joseph Broome, and Henry Samson. For Dinner Table Plants, T. M. Shuttleworth; G. Campbell, of Liverpool; and Joseph Broome. For Calceolarias, T. M. Lord, of Cheadle; and H. J. Leppoc, of Higher Broughton. For Pelargoniums, James Fildes, of Chorlton-cum-Hardy; and S. A. Meyer, of Prestwich. For Rhododendrons, John Heywood, of Stretford; and John Stevenson. For Marantas, T. H. Birley, of Pendleton; and Joseph Broome. For Cut Flowers, E. Wrigley; G. W. Binns, of Fallowfield; and H. T. Broadhurst, of Prestwich. In the Nurserymen's Classes, the successful competitors were:—For Orchids, B. S. Williams, of Holloway; R. S. Yates, of Sale; and William Rollisson & Sons, of Tooting. For Stove and Greenhouse Plants, E. Cole & Sons, of Withington; and Thomas Jackson & Son, of Kingston. For Azaleas, E. Cole & Sons; H. Lane & Son, of Berkhampstead; and Charles Turner, of Slough. For Rhododendrons, R. S. Yates; John Waterer & Sons, of Bagshot: and G. & W. Yates, of Market Place, Manchester. For Roses in Pots, Charles Turner and H. Lane & Son. For Alpine Plants, W. Rollisson & Sons: John Shaw, of Bowdon; and G. & W. Yates. For Hardy Trees and Shrubs, G. Caldwell & Sons, of Knutsford; and John Shaw. For Japanese Plants and Yuccas, John Shaw. For Hardy Ferns, Charles Rylance, of Ormskirk; and Thomas Jackson & Son. For New and Rare Plants, W. Rollisson & Sons and B. S. Williams. The Silver Cups offered by Mr. William Bull, of Chelsea, for New Plants of his own introduction, were awarded to Messrs. T. M. Shuttleworth, B. S. Williams, and E. Cole & Sons.

BRUCE FINDLAY, Curator.

BOTANICAL GARDENS, STRETFORD,
19th December 1874.

Findlay's printed report for the 1874 National Horticultural Exhibition (MBH 3/2/60).

Even though the weather on the first day of the Society's Whit Week Exhibition in 1869 had been dreadful, exclusivity drew 3000 visitors who had paid 10s 6d (in addition to members and their families) and 500 new subscribers had been recruited. Findlay's ambitions were being realised.

A Royal Crest and the words Royal Manchester Botanical and Horticultural Society head *The Report of the Council* for 1876, which states that:

Her Majesty the Queen has graciously consented to become the Patron of the Society, and on two occasions during the past year has also been an Exhibitor at the Society's Exhibitions held in the Town Hall.

Although commercial depression was again afflicting the cotton trade and membership levels had begun to fall, at the fiftieth AGM in 1877 the Chairman, Dr Watts, was pleased to say that there was only £2000 of debt left to pay and he trusted the end of the year would see its elimination. His hopes were realised and the Society was able to announce at the AGM in 1878 that the long term debts were eliminated and that:

> The £1000 owing to the Lancashire Insurance Company at the beginning of last year has been paid, and the Society is now in possession of the deeds of the estate. The £500 advanced to the Society thirty years ago by the late Edmund Buckley, Esq., has also been paid … The Society is thus relieved of all claims except the balance due to the Bank.

The continuing success of the National Exhibitions saw the Society solvent for the first time in its history. It was not to last long.

By 1879 the amount owing to the bank was rising again. As Whit week had been unprecedently wet the show had lost £500, which was doubly unfortunate as the Council had spent £900 on providing a new iron structure to cover the Exhibition Ground.

The Council noted other reasons for failure; a commercial depression in Manchester and a dense freezing fog that had enveloped the city, deterring visitors from attending the late show even though the Queen had sent an exhibit of 60 different kinds of apples. The fifty-third AGM in January 1880 saw the Council optimistically announce that the Society was back in profit: the shows had yielded £1032 12s. 10d. though they had only carried over a balance of £28 9s 3d Ominously subscriptions had fallen and 'Within a mile of the gardens there were 300 empty houses'. The Chairman had no doubt that people who had been their subscribers had tenanted half of these. These were worrying trends needing to be addressed. Never seemingly daunted Findlay and the Council announced in 1880 that, in 1881, their Jubilee year, in addition to the National Show in Whit week, there would be an International Show in August. Any surplus monies from these Exhibitions would go to replacing the glasshouses.

Horticultural Fate. From *Cruikshank's Humorous Illustrations* (London, 1890s).

The Manchester Society's reputation as a premier provider of exceptional horticultural exhibitions had never stood higher. The Whit Week Show was a resounding success, drawing 44,000 visitors and resulting in a profit of £700.[22] The *Gardener's Chronicle* described the International Exhibition in August as 'the greatest combined fruit and flower show ever held in this country'. Local recognition of Findlay's abilities followed almost immediately. A testimonial from the Society said: 'there is not a steward of the public welfare to whom the community is more indebted ... than to the excellent curator of the Botanical Gardens at Old Trafford'.[23] He was also presented with a cheque for £1000, a gold watch and chain and an address 'on vellum, morocco-bound'. To honour their own success, the Society's name was changed to *The Royal Botanical and Horticultural Society of Manchester and the Northern Counties*.[24] Findlay urgently pursued the construction of the new glasshouses fuelled by the knowledge that 1887 would be Queen Victoria's Jubilee year; a Jubilee Exhibition in Manchester would enhance the Society's reputation still further. With his past successes, the Council accepted his proposals without question and the £890 profit from the International Exhibition was put at Findlay's disposal. Findlay immediately presented his plan for an expensive semi-circular range of glasshouses. Even though the Society was now solvent, the cost of these plans was to stretch finances once again and echoes the expenditure prior to the Art-Treasures Exhibition.

To provide more income, a new 6d day ticket was introduced, admitting the general public at anytime other than Exhibition and Promenade days. Exclusivity had almost come to an end in the pursuit of new exhibition houses. In a reversal of the usual statements that the Gardens were private and only for proprietors, Findlay argued publicly that this was merely nominal as a Garden visited by more than 40,000 or 50,000 people a week was hardly private. Findlay was now in *de facto* charge of the Society and perhaps expressing his own beliefs publicly for the first time. By July the ranges had been replaced, with an energy that was seen as characteristic of the Society.[25] The following year Findlay unveiled yet more ambitious plans, although already the five new houses had added £1180 of debt. Findlay explained to the nation's gardeners :

> If the large exhibitions are to be continued a new house must be built. If the Whit-week exhibition was abandoned the united expression of many thousands of persons in that locality would be that their greatest pleasure during the holiday week was gone.[26]

The Society, he claimed, needed to construct a spacious glass building to hold the Society's valuable collection of tender plants or the week-long shows would be discontinued. In a February article on the subject, a reporter of the *Gardener's Chronicle* summed up the Council's attitudes succinctly, 'this is being done in true Manchester fashion – not too long thinking or talking about it, but onwards to its speedy completion'. By the end of April, the new house, 54 ft wide and 252 ft long [16.5 × 76.8 m], built by Halliday's of Middleton, was complete and ready for the National Horticultural Exhibition in May 1883.

The New Exhibition House and adjacent terrace and flower beds, Royal Botanical Gardens, Manchester (postcard n.d.).
AUTHOR'S COLLECTION

The New Showhouse (*Gardener's Chronicle*, 19 May 1885, 639).
© ROYAL BOTANIC GARDEN, EDINBURGH

Findlay, perhaps anticipating the cold, wet weather of past opening days, had gathered as many of the tender plants as possible under cover in the house. With this lavish building programme, it is not surprising that, at the fifty-eighth AGM, Findlay admitted they had spent £6000 over the past four years on glasshouses. Findlay's final work was to construct a new Palm House enabling tropical plants to be grown together. He argued this was an absolute necessity and, in February 1883, the Council sanctioned the house at a cost of £1000. When Findlay suggested that each of the 500 life members should enrol another subscriber at two guineas per annum to raise the money, the construction of the Palm House had already begun.

This was a return to old habits and at the 1887 AGM the Council conceded that the cost of the new Palm House had been £2000 and, in addition, there had been less income from the shows. They claimed major expenditure was completed and they assured proprietors that the Palm House would be a main attraction at the Manchester Jubilee Exhibition.

The members were more interested in whether there would be free entrance. The meeting was informed that Sir Joseph Lee for the Jubilee Executive Committee and Mr. Broome, the Treasurer, for the Council, had decided that:

Inside the Palm House, Manchester
Botanic Garden (postcard n.d.).
AUTHOR'S COLLECTION

A view of the Royal Jubilee
Exhibition including the Botanic
Garden and Olde Manchester and
Salford. From H. Garside, *The Royal
Jubilee Exhibition, Manchester: a
photographic record* (Manchester,
1877).
CHETHAM'S LIBRARY, MANCHESTER

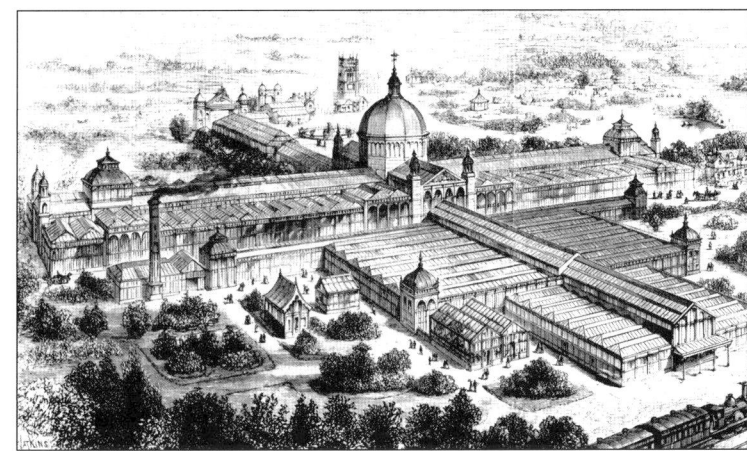

Plan of Old Manchester and Salford.
From A. Darbyshire, *A Book of Olde
Manchester and Salford* (Manchester,
1887).
AUTHOR'S COLLECTION

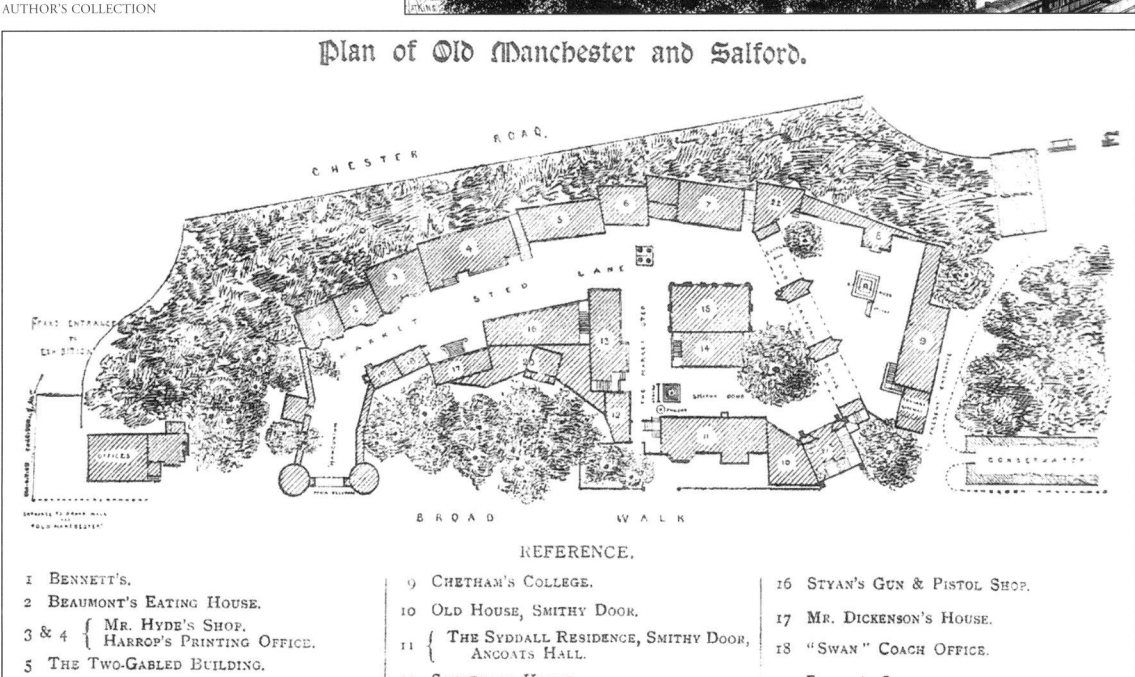

Plan of Old Manchester and Salford.

REFERENCE.

1 BENNETT'S.
2 BEAUMONT'S EATING HOUSE.
3 & 4 { MR. HYDE'S SHOP.
        HARROP'S PRINTING OFFICE.
5 THE TWO-GABLED BUILDING.
6 LOXHAM'S SHOP.
7 { NEWTON'S SHOP.
    "EAGLE & CHILD" COFFEE HOUSE.
8 { OLD SALFORD—THE RESIDENCE OF THE
    ALLENS.

9 CHETHAM'S COLLEGE.
10 OLD HOUSE, SMITHY DOOR.
11 { THE SYDDALL RESIDENCE, SMITHY DOOR,
     ANCOATS HALL.
12 SANCTUARY HOUSE.
13 OLD POST OFFICE & NEWSPAPER OFFICE.
14 BOWEN'S SHOP.
15 FIRST EXCHANGE.

16 STYAN'S GUN & PISTOL SHOP.
17 MR. DICKENSON'S HOUSE.
18 "SWAN" COACH OFFICE.
19 BARBER'S SHOP.
20 HULME HALL.
21 TOWER OF COLLEGIATE CHURCH.
22 OLD FIRE ENGINES.

Life Members, proprietors of, and subscribers to the Botanical Gardens would receive two Jubilee Exhibition tickets having the same privileges as a first-class season ticket issued by the Executive Jubilee Committee.

A motion was proposed that these arrangements were not satisfactory unless the season tickets issued allowed entrance 'on all occasions' and a large majority carried the proposal. Families were also to be given Botanical Society tickets to allow free entry to the four exhibitions in the Town Hall and the Whit Week Exhibition on the land at Old Trafford, rented for the purpose from the Deaf and Dumb School.

Manchester's celebration of the Queen's Jubilee in 1887, the Royal Jubilee Exhibition, was held on the same site as the Art Treasures Exhibition, and this

Montage of views and images of the Exhibition and Old Manchester and Salford a) the Roman Gate entrance to Old Manchester and Salford; b) one of the bandstands; c) A house in Olde Manchester and Salford; d) the Royal Entrance.

A) AND C) FROM A. DARBYSHIRE, *A BOOKE OF OLDE MANCHESTER AND SALFORD*, 89 (MANCHESTER, 1887), AUTHOR'S COLLECTION
B) POSTCARD (N.D.) AUTHOR'S COLLECTION
D) FROM H. GARSIDE, *THE ROYAL JUBILEE EXHIBITION, MANCHESTER: A PHOTOGRAPHIC RECORD*, CHETHAM'S LIBRARY, MANCHESTER

time the Council happily agreed to the Garden being incorporated into the site. One exhibit was indeed built in the Garden, the reproduction of Old Manchester and Salford.[27] Other changes to the Society's property included the large exhibition house turned into a dining room, the Annexe extended to form the Chester Road entrance to the Exhibition, grass transformed into terraced walks, and the huge fairy-fountain, built between the bandstand and the exhibition house. An additional bandstand was acquired and a commodious refreshment room built on the Trafford Road side of the Garden. The Prince and Princess of Wales opened the exhibition on 3 May and, after inspecting the exhibits, had lunch in the Palm House.[28] The question now exercising the Society was whether allowing the public such close contact with the Gardens would result in an increase in membership. Would the expense and debt be redeemed by their subscriptions?

Notes

1   'Royal Manchester Botanical and Horticultural', *Gardener's Chronicle,* 29 January 1881, 154.

2   The District Bank agreed in January 1866 to a new rate for the loan though the matter was adjourned. In July a letter from the Bank asked what was happening and the Council, with no explanation, decided to postpone their answer. Nothing more was minuted in 1866.

3   B. Elliott, *The Royal Horticultural Society: a history 1804–2004,* 120–1 (Chichester, 2004).

4   Letter, 'Manchester National Horticultural Exhibition', *Gardener's Chronicle*, 11 May 1867, 492.

5   For a history of the medals see W. L. Tjaden, 'The Medals of the Royal Horticultural Society', *Archives of Natural History* 21, 77–112 (London, 1994)

6   *Guardian*, 18 June 1831

7   This did not apply to all subscription botanic gardens. In September 1836, Sheffield advertised its shows nationally, had 11 judges and drew crowds of thousands. See 'Sheffield Horticultural and Floricultural Exhibition – open to all England', *Floriculture Magazine* 2, 116–20.

8   An Amateur, 'On Horticultural Societies and Exhibitions', *Floriculture Magazine* 4, 200–2.

9   Complaints about judges were frequent both locally and nationally. These included prizes awarded irrespective of merit, favoritism, lack of knowledge and inability to apply the rules and regulations.

10 *Liverpool Mercury*, 17 May 1816. The leader had been bandmaster to Napoleon Buonaparte.

11 'Notes of a trip to the Great Flower Show at Manchester', *Gardener,* July 1867, 251–7.

12 'The Schedule of the National Horticultural Exhibition', *Gardener's Chronicle*, 17 November 1866, 1089.

13 A. Pettigrew, Letter, *Gardener's Chronicle,* 12 January 1867, 30.

14 'The Manchester National Horticultural Exhibition', *Gardener's Chronicle* 21 December 1867, 1293.

15 'International Fruit Show', *Gardener,* August 1868, 360.

16 'Proposed subscription to Mr Bruce Findlay', *Memorandum, Manchester, 17 July 1867* (Manchester, 1867), MBH 3/2/55.

17 'Royal Horticultural Society's Great Provincial Show at Manchester', *Gardener,* September 1869, 422–33. Interestingly the Botanical Society did donate prizes for some of the floral classes, as did other local organisations, including the City of Manchester, the *Guardian*, the *Courier*, the *Examiner and Times*, local nurserymen and many individual proprietors of the Botanic Garden.

18 'Editorial', *Gardener's Chronicle,* 24 July 1869, 785.

19 Findlay had instituted some botanical beds to the rear of the glasshouses.

20 'Manchester Botanic Garden', *Journal of Horticulture and Cottage Gardener* 7, 1871, 439–40. The subject of pollution and the growth of plants in cities was a frequent topic in the gardening magazines throughout the nineteenth century. An article headed 'Town Trees for Manchester' in the *Gardener's Chronicle* on 22 October 1881 commented that within a radius of 20 miles of the city '… it is no exaggeration to say that there are scores – it may be hundreds – of acres of forest trees … which are now only to surely fast dying a premature death.'

21   A. Forsyth, 'Manchester Botanical and Horticultural May 26', *Gardener's Chronicle*, 3 June 1871, 713–4.

22   'Manchester Whit-week Show', *Gardener's Chronicle*, 11 June 1881, 794.

23   'Our Album Mr. Bruce Findlay', *Momus,* 25 August 1881 (Manchester, 1881).

24   'Royal Botanical and Horticultural Society of Manchester and the Northern Counties', *Gardener's Chronicle,* 27 January 1883, 121.

25   'New plant-houses at Old Trafford', *Gardener's Chronicle,* 18 May 1883, 639.

26   'Royal Botanical', *Gardener's Chronicle,* 27 January 1883, 121.

27   See A. Darbyshire, *The Booke of Olde Manchester and Salford,* 14 (Manchester, 1887). The organisers had been inspired by Exhibitions of 'Old London' and 'Old Edinburgh' that had recently taken place. Note the electric light. Electricity had been installed by 1888. Over the years suppliers changed, mainly due to cost and problems with supply were frequent. The Society considered buying its own plant but finally decided against it.

28   'The Manchester Botanical and Horticultural Society's Gardens', *Gardener's Chronicle,* 25 June 1887, 838. Suggestions were made when the Exhibition closed to keep the entrance Annexe, Old Manchester and Salford and the fountain as permanent features and to turn the refreshment room into a Winter Garden, but lack of finance prevented all of these.

CHAPTER TWELVE

'Our Poverty and not our Will consented'
1889–1894

Bandstand in the Botanical Gardens, Manchester from an article 'Opening of the Manchester Royal Jubilee Exhibition' *The Illustrated London News*, May 14, 1887.
AUTHOR'S COLLECTION

After the excitement of 1887, the Botanical Society had a commemoration of its own. 1889 was the 21st anniversary of the National Horticultural Exhibitions. The *Gardener's Chronicle* reported that:

> It will be a 'coming-of-age' celebration, and the Council is anxious to make it one of the most remarkable displays ever seen. ... The Society is now in possession of buildings for exhibition purposes not surpassed in Europe.[1]

The Society might have had the best exhibition buildings in Europe but the cost had once again strained their resources and financial problems loomed large. This time there was to be no reprieve. A description of the gardens in June 1888 showed that the Council had invested heavily in returning the grounds to their former state.[2] The enormous fairy fountain was considered too expensive to maintain and was broken up with the stone being used throughout the gardens.[3] Mr Clapham of Southport was employed to construct a piece of rockwork at the end of a new walk and he remodelled the crater where the fountain had stood as a sunken rock garden, which included new seating.[4] The Grand Avenue was reglazed and the ground plan reworked to produce an extensive promenade.

The Chester Road end of the Grand Avenue was converted into several rooms. The dining room at one end faced a spacious hall, said to seat as many as 1500 people, and at the other end, a platform was installed for an orchestra. Dressing, smoking and refreshment rooms were constructed along the sides of the hall. The lawns were returfed and the walks were resurfaced with white Derbyshire

Flower Show at the Royal Botanical Gardens, Old Trafford. Inside the Palm House. From W. A. Shaw, *Manchester Old and New* (Manchester, 1894).

spar. The publicity afforded the Garden by the previous year's Exhibition had increased the subscriptions by £3000 and as a result the Council decided that musical Promenades would increase to three times a week during the summer.[5] A bandstand had been bought for £60, thanks to the generosity of a member, Mr James Benton, who sent a cheque to cover the cost. By 12 September the subscriptions were up £3102 on the previous year. Disturbingly, Messrs Higginbotham reported that water, dripping from the higher roof of the Grand Avenue onto the lower, had washed away putty. This has echoes of the problems with Worthington's Exhibition House. Although there appeared to be sufficient money available, the Council left the repairs until the following Spring.

In 1888 the Council had let the catering for the year to a William Wood had to supply his own staff and equipment, the payment to the Council merely granting him access to the gardens. Wood's experience shows how dependent was the success of the Garden on the local residents for, although the Garden was accessible by rail and omnibus, the bulk of the members visiting regularly lived locally. In November 1888 Wood informed the Council that his season had not been satisfactory.[6] Expenditure had been £3007 (not including his time and labour) and his receipts were £2920. His success the previous year at the Jubilee Exhibition, he claimed, had misled him but in reality even in a good season, he would see little profit because:

> The people who have frequented the gardens reside in the immediate neighbourhood, and consequently have little need for refreshments. … The gardens have only been open for 69 days and this means nearly £50 per week of 6 days for *rent alone*, and there is only a few hours per day in which to do my business.

A report in the *Manchester Guardian* as early as 1862 had correctly predicted the consequences for the Society of the move to the suburbs, claiming that as 'gentlemen had gardens of their own, and only cared to go on exhibition days or for the occasional promenade for a small fee' membership would decline.

Even with improved transport, many possible subscribers were deterred from visiting the Garden – to Wood's and the Society's detriment.[7]

The Council realised that even with the recent increase in membership the finances needed addressing and a special fund was started again, confirmed by a letter from Balmoral:

> Queen's Secretary Balmoral Castle 12 September 1889
>
> Sir, Having laid your letter before the Queen, I am commanded by Her Majesty to inform you that she will give £25.00 to the Royal Botanical Garden, Manchester.
>
> Yours faithfully, Henry F. Ponsonby[8]

The same minutes record a letter from Lord Derby agreeing to attend a special public meeting to discuss the rescue and an account of the meeting showed 150 persons were present and all wanted the horticultural and musical events to continue.[9] The Council pledged to use every means possible to satisfy the members and in pursuit of this aim started to charge for entrance to the Concert Hall. In the hope of attracting new subscribers, the Garden was opened free on 23 November; 13,000 people entered the grounds.[10] Though subscription monies were £3328 in July 1889, they were down £1608 on the previous year as the boost given to the membership by the Jubilee Exhibition began to fade. By June 1890 the money from subscribers was down a further £2000 on the previous year and the gate receipts for the Whit Week Exhibitions were down by £206.[11]

The newly revamped Garden needed a considerable increase in finances if it was to be maintained. In the years for which the minutes are missing, the Council had apparently been attempting to gain income from letting the Garden, although free admission for Society members and their families to any event was taken for granted. Again and again in later years it becomes clear that this free admission policy was to cause problems when attempting to let the gardens. Potential hirers quickly realised that their projected profits, if based on previous gate numbers, were in fact a fiction. From 1890, Council minutes of special meetings to consider applications for the use of the Garden, show how lettings were organised. For example the Freemasons of East Lancashire Garden Party was held on 26 July and the Council's conditions were that all subscribers and their families were to have free admission and the minimum charge for the public was 2s 6d per adult and 1s per child, presumably to maintain exclusivity.[12] After expenses the Society was to have 50% of the sale of tickets and gate receipts and the Council reserved the right to approve any exhibitions or amusements. Notwithstanding the Garden continued to be an attractive venue to outside organisations.

In 1892 an offer was received that had seemed to solve all the Council's problems; Mr Mumford of London wanted to use the gardens for an Exhibition of American Ancient Cities.[13] The 'Old America' show included the reconstruction of three American towns, a 170 ft (nearly 52 m) replica of the Eddystone lighthouse, a reproduction of Columbus's ship *Santa Maria* and Blondin walking the tightrope. The Council agreed, no doubt expecting the venture would draw the same crowds as the exhibition of Old Manchester and Salford; in this they were to be disappointed. Not only did the Society lose

£2250 from the exhibition but also the cancellation of their own major shows meant they had little other income; some small horticultural shows did take place, both at the Town Hall and the Garden. As a small nod to the financial situation only hereditary members and their families were to have free entry, annual subscribers were to pay. The terms for the hirers were strictly commercial: the Garden was to be left in good condition and any damage repaired and water and refreshments were to be paid for. The Palm House, Dining Room, Grand Avenue, Concert Room and adjoining anteroom could be used but the Exhibitors would meet any alteration costs.[14] Considering the value of these buildings, the Council's surprising willingness to agree to any alterations at all is perhaps a measure of their desperation to settle their debts.

A special meeting of subscribers voted to accept the offer, and the Society proposed fee of £4250 was agreed.[15] £1000 was to be paid when the contract was signed, £1000 on 1 June and 1 September and the balance on 31 October. Another special meeting of proprietors was held on 14 March 1892 agreed to allow the Garden to be closed from 20 March to the end of the year. By October the Old America Company owed £1000 and on 11 November they had gone into liquidation and £250 was offered as a final settlement. The Council accepted, although the £250 was not received immediately and the Council applied to the liquidator who forwarded it on 2 August 1893. Despite the difficulties, in 1894 the Committee, again acknowledging Findlay's importance as the man who had proved central to the Society's survival and they presented him with 'a handsome silver service and Mrs. Findlay received a gold ring'.[16] Findlay's reply to the honour was sobering. He was unhappy about the use the Garden was being put to and felt they now lived in an age of 'sensationalism and frivolity'. Soon he had other matters to concern him. Like Campbell before him, he was to suffer accusations of professional negligence.

The dispute was again, as in 1857, to reveal the inner workings of the Society and show that Findlay's influence may not have been as positive as was publicly portrayed. A series of three letters was the start of the dispute. The author was Benjamin Armitage JP, a former member of the Council, Chairman of textile manufacturers, Sir Elkanah Armitage and Sons Ltd, who had served as the Liberal MP for Salford from 1880 to 1885.[17] To discuss the letters' allegations, a special

Mr B. Armitage MP
From W. T. Pike, *A Dictionary of Edwardian Biography*, (Brighton, 1899, reprint Edinburgh, 1987).

Mr. B. Armitage, J.P.

Armitage.—BENJAMIN ARMITAGE, J.P., Chomlea, Pendleton. Second son of the late Sir Elkanah Armitage, of Hope Hall, Pendleton. Born, 1823; Chairman of Sir Elkanah Armitage and Sons, Ltd., manufacturers; Treasurer of the Manchester Chamber of Commerce; Director since 1870; President, 1878-81; Justice of the Peace for the city of Manchester and the county borough and hundred of Salford; represented Salford in Parliament as a Liberal, 1880-85; represented the West Division of Salford in the short Parliament, 1885-86; was the political supporter and intimate friend of Richard Cobden and John Bright; co-optative Governor of Manchester Grammar School; has taken special interest in extending the operations of the Sick Poor Nursing Institution, for the joint constituencies of Manchester and Salford. On January 18th, 1899, received the honour of the Freedom of the County Borough of Salford.

Council meeting immediately took place with Findlay present. He had taken exception to the third letter and said that in the opinion of his solicitors, the letter was libellous. Findlay admitted that for some time he had been 'subjected to petty persecutions' from Armitage and that he was resolved to take legal proceedings. On 25 May Findlay's solicitors sent a letter to Armitage asking him to withdraw his comments. In a letter dated the next day Armitage declined:

> If my letter on the present condition of the Botanical Gardens had pointed directly to your client and was intended as you write to be libellous, untruthful, and malicious, I should not have been the author of it ... I must reject your client's gratuitous demand to apologise to him for my letter of which he so unjustly complains.[18]

Armitage had, in fact, written three letters to the *City News*, and it was the third that provoked Findlay to action. The first letter complained about the debt ridden-state of the Society and said that the Garden had 'been in pawn for fifty years', an exaggeration but, as the evidence has shown, with more than an element of truth.[19] In the second, Armitage, who claimed 17 years experience on the Council, admitted the expenditure had been extravagant and suggested that the Garden should be open every Sunday 'as the Whitworth Park is every day' and considered this would keep the Garden open 'before it closes its doors with a millstone of debt round its neck'.[20] Again the history of the Society has shown extravagance by the Council had led to debts but Sunday opening in 1870 had not solved the problems, and allowing the general public in on Sundays at this point might do no better.

Armitage perceived that there was a greater problem namely, the management of the Society. He complained that Findlay was 'acting as a dictator, secretary, musical director, treasurer, and "boss of the whole show"'. How far this statement was true is hard to gauge; perhaps it was a question of perspective. Certainly, Findlay was all the things Armitage listed regarding the management of the Society, bar possibly dictator. However, as the Council had presented a tribute to the Curator in the very week Armitage wrote his first letter, they cannot have seen this as sinister. On the other hand Findlay can be seen to have been the originator of the grandiose glasshouses that had brought about the current debts and it could be argued he may have acted in a dictatorial fashion in pursuit of this. Findlay chose to ignore these comments. It seemed that Armitage's dispute was with the Council but testimony at the later trial showed there was a long-standing dispute between himself and Findlay.[21] The trouble started after the wedding of Findlay's daughter in September 1887 and involved a disputed payment for photographs.[22] Armitage had admired a photograph of himself, ordered several copies and sent the bill to Findlay, who declined to pay. Armitage then started sending Findlay abusive letters and arguing with him in Council meetings. Armitage became a Council member in 1876 but evidence at the trial showed he had been 'quietly shunted from the Council' in January 1894 as the other Council members found him difficult.[23]

The third letter was of a different character. In it Armitage listed complaints about the Whit Week Show and the running of the Garden. The first letter can

Frivolous pursuits:
the Belle Vue dancing
platform (postcard, n.d).
AUTHOR'S COLLECTION

be related to the move of the wealthier members to the suburbs and also to the commercialisation due to the Society's attempts to obtain cash. The second letter, if true, could be a direct result of the financial difficulties of the Society. Findlay took this third letter as a personal attack. First Armitage claimed that the first day of the show was 'inferior' as the amateur classes were no longer attracting the gentlemen of the neighbourhood, and their place had been taken by 'the trade' who sent their exhibits to sell.[24] Further, there were no exhibits from the Society – the expensive glasshouses had produced nothing worth showing. He claimed that, having visited the nursery at the Garden, he had seen orchids dying from lack of proper care – the same accusations levelled at Campbell in 1857. Bankruptcy, he declared, was inevitable. Although Armitage claimed in court that he was accusing the Council of dereliction of their duties, Findlay took it personally.

For the Whit Week Show at least, there was a seemingly independent witness, the author of the *Gardener's Chronicle* report published on 19 May.[25] The reporter called it 'imposing' although noting that the 'elephants' of the past, the miscellaneous stove plants and the orchids, were absent. This suggests that Armitage was correct in maintaining that the elite middle class was no longer exhibiting. The pot roses, mainly a class for professional growers, were described as outstanding. These were an example of the trade exhibits Armitage complained about. The orchids in the Exhibition House, he continued, made a magnificent display but 'more one of general purposes than one showing ten or twenty years cultural skill' as had been the case in the past. Again, Armitage seems to have been at least partially correct about the glasshouses though his animosity towards Findlay seems to have led him to exaggerate on this point. The reporter's main complaint was related to the exclusivity of the Manchester shows. Although the opening day was fine, the select 'upper ten' of the local aristocracy had not come in such numbers as he would have liked or expected to have seen at such an important show. He attributed their absence in part to the death of Lord Derby's sister, which meant Lord Derby had had to send his apologies for his absence.[26] For the upper echelons of the gardening public seeing the aristocracy patronising national shows would seem to have been as important in 1887 as it had been in 1827. The desire to attract the expected aristocracy may have been another factor in the determination of the Manchester Society to remain exclusive on the opening day.

Letters in the *City News* supported Armitage's view. On 2 June 'Looker-on' said the Society was not above criticism as it raised money from 'handing round the hat' and called for 'cheap entrance, railway excursions for the millions outside Manchester to see the "vast treasures of nature"'.[27] On 9 June a second correspondent, 'An Old Subscriber', saw the Garden as declining and the tastes for recreations changing: a remark reinforcing Findlay's comments on frivolity.[28] If the Society wanted support, he asserted, it would have to change from 'a middle-class recreation ground' as there were several free beautiful public gardens in Manchester. Further, he declared, financial expenditure needed curbing, and plants and fruit should be sold. Southport Botanic Garden Society, he noted, paid proprietors a 4½% dividend and encouraged visitors by opening the garden every day for 4d.[29] The Society, he claimed, was 'in the sorrowful position of many old established houses ruined for want of spirit to adapt themselves to change'. This seems a perceptive comment on the expectations and actions of the Council. In reply to these letters, Armitage announced that the libel trial of Findlay v Armitage was to take place on 12 July 1894 and hoped that numerous readers would delay their holidays and attend.

Rather than genteel exclusivity, the evidence from the trial paints a very different picture from the one promoted by the Council and confirms that it was engaged in a variety of commercial enterprises to try and raise capital. The case gives clear evidence of the lengths the Council was willing to go to in order to liquidate the debt. The activities they had promoted bore little relationship to horticulture and had in fact made their financial position worse. Findlay's testimony, and his speech at the Whit Week presentation, made clear that he did not personally support the activities the Council had promoted. As he commented at the trial, 'Our poverty and not our will consented'.[30]

At the trial Findlay stated his belief that the Garden provided benefit to the community at large, showing that in 1894 the Society had appealed for public support on the basis that they provided 'public services for public welfare'. Findlay's evidence gave his personal view of the situation of the Garden and the debts. While admitting that the proximity of the town had an effect, he stated that the gardens were well looked after and, though botanic gardens were not unique to Manchester, the Manchester gardens were as good as those found elsewhere. He had told the court that as Secretary, he was aware of the financial situation and, as the value of the property had greatly increased, there was no suggestion that the Society faced bankruptcy. This seems rather disingenuous of Findlay as in 1890, the Society had an acknowledged debt of £6000 and a public meeting had been called to start a debt fund. It also ignores the obvious fact that, if the debts became too great, the only solution would be to take advantage of the price of the land and sell the Garden, thus destroying the very thing he was attempting to preserve.

Mr Shee, counsel for the defence, painted a jaundiced view of the Council as self-serving commercial businessmen with members of the Council supplying

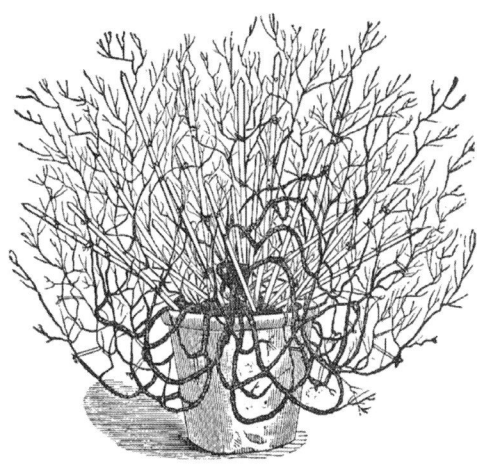

SKELETON DWARF TRAINED CHRYSANTHEMUM.

DWARF TRAINED CHRYSANTHEMUM.

Left: dwarf trained chrysanthemum. Those who cultivate Chrysanthemums for exhibition commence operations by striking the cuttings in November and December. In the spring they shift the plants and encourage growth in a frame or pit; *Right*: dwarf trained chrysanthemum. From S. Hibberd, *The Amateur's Flower Garden*, 271–2 (London, 1884).
AUTHOR'S COLLECTION

their commodities to the Society. This is hardly important but what it does reveal is that the nature of the Council members had changed; they were no longer mainly drawn from the elite middle class but from Manchester's commercial community. Though the Society claimed its principal object was the promotion of the science and art of horticulture, Shee listed other activities they had promoted: cat shows, dog shows, bands and 'Old America'.

Findlay was forced to agree that it was not a way to run a Botanic Garden. When asked if the gardens were successful, Findlay replied that intellectual pursuits were not appreciated as 'frivolity has swept the nation'; a view he had held for some time. He also conceded that the number of subscribers was falling but 'that every institution in the country is complaining of the same thing'.[31] In 1891 this was true, for example, of Manchester's private libraries; the Portico Library was in severe financial trouble, as were the Exchange and Foreign Libraries; the latter two closed in the early years of the twentieth century.[32] It was also true of other subscription botanic gardens as they confronted similar difficulties with varying degrees of success. In his summing up the Judge said the letter could be construed as 'fair comment'. The jury disagreed for, after retiring for a short absence, they found for the plaintiff, awarding £25. In the second action, against the *City News,* they awarded Findlay £35.

Findlay, unlike Campbell, survived the crisis and it must have been of comfort that the Autumn Show held at the Town Hall on 23 and 24 November was said 'to be the finest ever held in Manchester … the chrysanthemums, supported by a display of beautiful orchids, drew an attendance much greater than we have seen for years'.[33] Perhaps this was a display of public support for Findlay and the Society or, just as likely, the idle curiosity of the general public.

Notes

1   'Royal Botanical and Horticultural Society of Manchester', *Gardener's Chronicle,* 3 March 1888, 274.

2   'The Botanical Gardens, Old Trafford', *Gardener's Chronicle,* 16 June 1888, 744.

3   Letter to the Editor, 'An Old Subscriber', *City News,* 9 June 1894.

4   Clapham had used as the base for rockwork a form of concrete of his own invention, Claphamite. Perhaps this was a Northern rival to Pulhamite. See S. Festing, 'Pulham has done his work well', *Journal of the Garden History Society* 1984, 12(2), 138–9; 'Notes'. *ibid,* 1997, 25(2), 230–7.

5   'The Botanical Gardens, Old Trafford', *Gardener's Chronicle,* 16 June 1888, 744.

6   Letter, 7 November 1888, William Wood to the Council, Minutes, 13 November 1888, MBH 2/1/5.

7   'Manchester Botanical Garden', *Guardian,* 16 January 1862.

8   Letter, H. F. Ponsonby to the Council, Minutes, 9 October 1889, MBH 2/1/5.

9   Report of Public Meeting, Minutes, 5 November 1889, MBH 2/1/5.

10   Minutes, 11 December 1889, MBH 2/1/5.

11   Minutes, 11 June 1890, MBH 2/1/5.

12   Minutes, 20 June 1890, MBH 2/1/5.

13   Minutes, 2 February 1892, MBH 2/1/5.

14   Minutes, 10 February 1892, MBH 2/1/5.

15   Minutes, 18 February 1892, MBH 2/1/5.

16   'Presentation to Mr. Bruce Findlay', *Gardener's Chronicle,* 19 May 1894, 632–3.

17   For Benjamin Armitage see W. B. Tracy and W. T. Pike, *Manchester and Salford at the Close of the Nineteenth century: contemporary biographies* (Brighton, 1899).

18   Letter, B. Armitage to the editor, *City New,* 26 May 1894.

19   Letter, B. Armitage to the editor, *City News,* 3 February 1894.

20   Letter, B. Armitage to the editor, *City News,* 17 February 1894.

21   'The Botanical Gardens Actions for Libel Two thousand pounds damages claimed: Twenty-seven pounds awarded', *City News,* 14 July 1894.

22   'Manchester Marriages', *Gardener's Chronicle',* 10 September 1887, 312.

23   *Ibid.,* 'Actions for Libel'. There are no minutes for the period in 1887 when the dispute began and the minutes from 1888 onwards do not allude to any letters about, or complaints relating to, Findlay.

24   Letter, B. Armitage to the editor, *City News,* 19 May 1894.

25   'Manchester Royal Botanic', *Gardener's Chronicle,* 19 May 1894, 633–5.

26   Minutes, 11 April 1894, MBH 2/1/5.

27   Letter, 'Looker-on' to the editor, *City News,* 2 June 1894.

28   Letter, 'An Old Subscriber' to the editor, *City News,* 9 June 1894.

29   See 'New Botanical Gardens at Churchtown, Lancashire', *Gardener's Chronicle,* 21 August 1875. Southport Botanic Garden, founded in 1874, was opened at Churchtown in 1875. The Society became bankrupt in 1935. The gardens were re-opened by the Council in 1939 and are extant.

30   Editorial, 'The Botanic Garden Libel Action', *City News,* 14 July 1894. The quote comes originally from *Romeo and Juliet,* Act One, Scene 5 as 'My Poverty, not my will, consented … .'

31   'The Botanic Garden Libel Action', *City News,* 14 July 1894.

32   A. Brooks and B. Haworth, *Portico Library: a history,* chap. 8 (Manchester, 2000).

33   'Manchester Royal Botanic', *Gardener's Chronicle,* 1 December 1894, 672.

CHAPTER THIRTEEN

Struggling On 1895–1907

Though Findlay had been vindicated, the Garden was causing anxiety. On 14 November, 1894, Findlay reported that the large glass partition in the Palm House was dangerous.[1] Major repairs meant there was to be no diminution of the debt. However in the city, Findlay had had a personal success. His devotion to the spread of horticulture enabled him to see a way of encouraging its practice locally and, with the support of the Council, 'to render service to the community'.[2] In a letter from Findlay to the readers of the *Journal of Horticulture and Cottage Gardener,* he explained that the Council proposed to hold an annual exhibition:

> ... on a very large scale of the productions grown by tenants of small holdings, at which prizes will be awarded for the fruits, flowers, vegetables, poultry, cheese and butter ... this special effort will be the means of stimulating the poorer classes of Society in their endeavours ... The Council is of the opinion ... it is the proper and legitimate work of the Society to introduce this project.[3]

Findlay noted that latterly many small tenanted allotments had sprung up around large towns. What better scheme could have been devised to counter the claims made at the trial that the Society did not fulfil its proclaimed purposes of encouraging horticulture?[4] The finances connected to the scheme were as problematic as usual and raises questions yet again of the financial acumen of a Council already heavily in debt. The Council had estimated the annual outlay at £2000 and a special fund was being established, this time for £5000 for this 'beneficent purpose'. Findlay had already written to the Queen and obtained her support in the form of a cheque for £25 and Findlay was 'pleased to state that £1000 had already been promised and the names of the donors would be published shortly'.[5] The editor of the *Journal of Horticulture* described it as 'an admirable undertaking and Mr. Findlay was to be congratulated on the excellent beginning'.

The last Council meeting of the year took place on 11 December 1895, when the dates for the next year's shows were agreed, the allotment show was to be held on the 17 September 1896. The Marquis of Lorne, the son of the 8th Duke of Argyll and MP for South Manchester, was to be asked to open the Whit Week Exhibition and Rt Hon. J. A. Balfour MP for the Eastern Division of Manchester, who had become Prime Minister in June 1895, to open the Rose Show.[6] At the same meeting it was agreed to send a letter of condolence to Findlay on the death of his wife. There were two proposed innovations for 1896,

Inside the Fernery,
Royal Botanic Gardens,
Manchester (postcard,
n.d.). Note the tree ferns.
Dickensonia were named
after James Dickson,
a famous British
cryptogrammic
botanist. *D. antartica* was
introduced in 1786.
AUTHOR'S COLLECTION

which attempted to generate an increase in membership. First, the restoration of the grand lawn meant that the Promenade could take place there as in the past.[7] Secondly, lady cyclists were to be allowed to use a special track, 'the broad walk on the Talbot Road side of the lake' though they could not introduce strangers.

Disaster overtook the Society when Findlay died suddenly at the Garden on 16 June. Committees were rearranged to take care of his duties. On 1 July 1896 the Council sent their condolences to his family and organised the Rose Show; the Aparian Society and Allotment Shows were abandoned. The question of the new Curator was deferred.[8] Obituaries to Findlay were published in all the national gardening magazines as well as in the local press. Several mentioned that his health had been failing, and Manchester's *City New's* obituary attributed this to the recent death of his wife. *The Garden,* a Scottish gardening magazine, claimed in their obituary, 4 September 1896, that an appeal for his daughters by Findlay's Manchester friends raised over £1200.

Cycle track at the Manchester Botanic Garden (postcard, n.d.). The lady members could not cycle on Flower Show or Promenade Days (*Minutes,* 8 January 1896, MBH 2/1/5). Physical activity was considered healthy exercise: B. Haley, *The Healthy Body and Victorian Culture* (Cambridge, Mass., 1978).

Royal Botanical and Horticultural Society
of Manchester and the Northern Counties.

PATRON - HER MOST GRACIOUS MAJESTY THE QUEEN.

Arrangements for the Year 1896.

✦ ✦ ✦

The Season will open at the Gardens on **Good Friday** with a MILITARY BAND from 3 till 6 p.m., and a GRAND DISPLAY of SPRING FLOWERS.

Easter Monday.

MILITARY BAND from 3 till 6, and from 3-30 Mr. JOSEF CANTOR'S "GEMS OF THE OPERAS COMPANY."

April 25th.

The National Auricula Society's Exhibition, with Mr. G. W. LANE's CONCERT commencing at 7-30.

From this date the Season will be continuous every Saturday evening. Band commencing at 6 o'clock. Mr. G. W. LANE's Concert at 7-30.

The Grand National Horticultural Exhibition,

which has been for so many years a source of the purest delight to many thousands in the locality, will open on the Thursday before Whit Sunday instead of Friday. The arrangements in connection with the Opening Ceremony will shortly be announced.

The Great Rose Show of the Season will be opened on the 25th of July by The Very Rev. The DEAN OF ROCHESTER.

The Special Exhibition in connection with the **Allotment Scheme** will be held on the 11th and 12th of September.

The Chrysanthemum Show will be held in the Town Hall on the 20th and 21st November.

Musical Arrangements for the Season.

The Band will commence every Saturday evening at 6 o'clock.

Mr. G. W. LANE's Concert will commence every Saturday evening at 7-30.

Conveyance Accommodation.

The Carriage Company will (by way of experiment) run Conveyances every Saturday on and after the 25th of April, commencing at 6 p.m., to the Gardens, from "The Green," Chorlton-cum-Hardy, "The Britannia," Whalley Range, and Victoria Park entrance, the end of Moss Lane East.

Arrangements for the year 1896.

The Council finally faced the problem of appointing a new Curator and Secretary in October and placed advertisements for the post in the national gardening press. On 28 October 1896 the minutes listed four candidates: Mr John Knight, London; Mr W. G. Baker, Botanic Gardens, Oxford; Mr P. Weathers of Silverhall Nursery, Isleworth; and Mr C. Paul, foreman of the Manchester Garden. Though the candidates had considerable experience, they were not of the calibre of Findlay. Weathers was appointed and started on 1 January 1897. At the same meeting, the Council accepted the resignation of the Chairman, Joseph Broome, as he now lived in north Wales and was finding travelling difficult, though the minutes of 3 February showed Broome was re-elected as Chairman.[9] Many financial problems remained to be dealt with and to these was added the Allotment Fund started by Findlay. The Council decided to cancel the related show and to refund cheques if requested. In July, perhaps as a final gesture to the concept of exclusivity, they had 'Royal Botanical Gardens' lettered over the entrance.[10]

By 6 October the Bank had replied to a request for an increase in the overdraft, saying that, on the contrary, the Bank wanted it reduced. On 10 November

Gates to the Botanical Gardens, Old Trafford (postcard, n.d.).
AUTHOR'S COLLECTION

Horsedrawn tram as used for transporting members of the public to the gardens (postcard, n.d.).
AUTHOR'S COLLECTION

An Aside: Patrick Weathers, Curator 1897–1907[1]

Patrick Weathers was born in Ireland in 1869. Described as a man of brilliant promise on his appointment as Curator of the Manchester Botanical Garden, Weathers had had a variety of horticultural posts before his appointment. After a few years of varied horticultural training, he moved to Silverhall Nurseries, Isleworth, before enrolling at Kew in 1885. After completing their training course, he was employed in the Assistant Curator's department. He left Kew in 1889 to return to the nursery trade where he worked for F. Sanders and Co., St Albans, who specialised in orchids. He moved to another orchid specialist, W. L. Lewis and Co., Southgate, before becoming the United Kingdom representative for the Belgian firm of Messrs Linden, probably the most famous orchid firm of the late nineteenth century. He then set up on his own, purchasing the Silverhall Nursery, where he rapidly expanded the business and established an orchid exchange scheme.

In 1897, he was appointed Curator and Secretary of the Royal Botanical and Horticultural

The Swiss Cottage in the Manchester Botanic Garden (postcard, n.d.).
AUTHOR'S COLLECTION

Society of Manchester and the Northern Counties. When the Garden was leased to the White City Co. Ltd Weathers became Secretary to the Society, though he continued to supervise the Garden.[2] He seems to have left Manchester around 1923 and returned to Isleworth. Weathers had a famous elder brother, John, who had a distinguished horticultural career and in later years published horticultural books and was Garden Editor of *The Field*. When his brother died, Patrick Weathers took over this position. Weathers was described a charming and amusing man who brought his clever appreciation of horticulture to all his endeavours. He died in January 1933 and was survived by a widow and five daughters.[3]

1   Obituary, *Gardener's Magazine*, 9 January 1897, 17. Information is from this source unless specified.
2   Obituary, Patrick Weathers, *Gardener's Chronicle*, 11 February 1933, 108.
3   Obituary, *Journal of the Kew Guild* 1933, 5(40), 278.
ALL © ROYAL BOTANIC GARDEN, EDINBURGH

the Council met with the Trustees to consider their response. The members were split over what to do. Some thought 'the gardens have had their day' and wanted 'no further part in spending the Society's money'. Others were more positive and said 'a new future was in store with many more subscribers leading to the gardens being self-supporting.' Mr Bowden concurred but wondered if they could sell part of the land. The Society's solicitors, in attendance, could not give an immediate answer. The meeting was adjourned until 15 December when the Council voted to send the deeds to the Bank as collateral. This was confirmed at a Special Meeting of subscribers, and the bank then authorised a maximum overdraft of £10,000 to be drawn without further authority.[11] At the same meeting Broome resigned again. A letter was read from Mr Tait, a Council member and retired nurseryman, of Llanfairfechan, north Wales. He suggested they reduced subscriptions and entrance fees as a way to solve the fall in members. The Council adopted the suggestion immediately. In May 1000 complimentary tickets were sent to 'influential Mancunians' to visit the first day of the Whit Week exhibition. This stratagem had been tried several times before, never it seemed with success; the costs of printing and distribution were never recorded or the return they might have expected. By December, Mr Tait had resigned as he was rarely in Manchester. The deficit for the year was £600 and there was £340 interest to pay to the bank.[12] As 1899 opened the Council had determined to sell plants and flowers to the public, only to see their plans thwarted by opposition from subscribers who were members of the trade. A compromise was reached, the trade had no objection to the Society growing tomatoes and Weathers was instructed to plant them.[13]

More problems beset the Society. In 1893 Humphrey de Trafford had offered Trafford Park to Manchester Corporation for a public park but they turned it down.[14] Now his heir was keen to sell the land for development and it had come to the attention of the Council that Royce Ltd, electrical engineers, had bought land in Trafford Park, opposite the Garden, to build an electrical works. This appeared to be in breach of the covenants to the Society within the original conveyance and they asked their solicitor to investigate. A second commercial development came to their notice: works proposed by Kilvert and Sir W. H. Bailey within 400 yd (366 m) of the Garden (see below), and the solicitors were asked to serve notices with reference to the Society's covenants on both developers.[15] The correspondent of *Atlantic Monthy*, Charles Elliott Norton when looking back towards Manchester when visiting the Art-Treasures Exhibition in 1857 said:

> Two miles off lies the body of the great workshop-city, already stretching its
> Begrimed arms in the direction of the Exhibition. … where Nature, reluctant
> To be driven utterly away, still tries to keep a foothold, she is parched and
> Scorched by the feverish breath of forges and furnaces. Standing here one may
> See the cloud of smoke, which waves in the wind like a pall over the city slowly
> moving and settling over the land.[16]

Now the 'begrimed arms' were on their doorstep, would members relish the prospect?

The decline of the Manchester Society was inevitable; an industrial estate was being built within sight of the Garden and must have contributed to the fall in subscriber numbers and the decline of visitors to events. Even though the Council had to admit to the financial implications, new ideas were attempted; an organist was hired rather than the bands of the past, and a garden party

Cotton Factories, Union Street Manchester. From S. Austin, J. Harwood, G. and C. Pine, *Lancashire Illustrated* (London, 1832). Manchester's 'begrimed arms' already there in 1832.
AUTHOR'S COLLECTION

was organised for the funds, raising £69, and the proposed sale of a parcel of land to the Deaf and Dumb School might offer a temporary relief.[17] But the amounts involved were trifling compared to the debt. A matter resolved in 1900 was the question of the restrictive covenants; the Society had the power, under their original covenants, to prevent certain works being built within a radius of 400 yd of the Garden. In the event, the Trustees met on 26 June and arranged to meet a deputation of the Directors of the Trafford Park Estate Company who wished to 'come to terms with the Society on the matters at issue'.[18] The Company offered £250 and the 'law costs' if the Society would give a release from the covenants. The Society was prepared to entertain the offer if a similar release was granted to the Society from the covenants that bound them.[19]

1901 saw the Society in profit by £188 6s 6d, but with hindsight it is easy to see that it was hardly possible the Garden could be maintained and attract

The ocean waves, Belle Vue (postcard, n.d.)
AUTHOR'S COLLECTION

subscribers as an industrial estate grew around them. One can speculate on the outcome for the history of the Garden if the Council had applied the power under their covenants. Would there have been a different outcome? It seems unlikely. To the Council, however, it was a new beginning in a new reign. The Council reviewed the state of the buildings and found that they were in a very bad state of repair.[20] More importantly, the following year the Council decided to discontinue the Whit Week shows on the grounds of economy from 1902, though this was later rescinded.

The finances were very much on the edge as 1902 progressed, the Whit Week show making a profit of just £24 12s. A Mr Gorer wrote in October requesting to use the gardens for an Industrial Exhibition from December 1902 to mid-January 1903 and the Council agreed. On 16 December the Council was happy to hear that Mr Gorer's second instalment of £150 was in the bank. 1903 opened with the bad news that he had reneged on his final payment and their solicitors advised they could recover the £100 by levying distress on any of his goods in the Garden. The Council decided to give him the relief of £100 and their generosity brought the loss for 1902 to £800 rather than £700.[21]

Finally, in 1903, the Council took the decision to sell the Garden to liquidate the debts, first to Stretford District Council and, when that failed, to Manchester Corporation.

Although the sales were unsuccessful for different reasons, the failure of both can be directly attributed to the financial problems experienced by the Society in struggling with the considerable capital costs of operating the Garden. On 12 February 1903, it was suggested that the Council ask Manchester Corporation to take over the gardens. On 12 March, before Manchester had been approached, a letter was read from Stretford District Council (hereafter SDC) informing the Council that they wished to purchase the Garden and enquired what their terms would be. The special sub-committee met with SDC at the beginning of May and SDC agreed to submit their proposals in writing, which were then discussed on 2 July.

The Council considered the terms fair but wanted to build a Horticultural Hall which would remain in its possession. The SDC offer was considered too low by several members, whilst others thought it too generous, but the Council finally agreed to accept an offer of £15,000 plus an annual subsidy to the Society. They then heard unofficially from SDC that it wanted to commute the annual subsidy to a lump sum *in lieu*. After much negotiation the Council met on 12 November and agreed to accept, subject to the approval of the Trustees. On 11 December it 'was mutually agreed' that £20,000 would be paid by the SDC for the land and the buildings and the SDC would put the buildings in 'a substantial state of repair' and maintain the gardens in a satisfactory condition. The SDC specifically agreed to keep in good repair those buildings needed by the Society for the flower shows, the Exhibition House, Grand Avenue, Concert Room and Refreshment Room. The Society would be given exclusive use of the gardens and buildings for 15 days each year and could employ their own caterer on those days. For the next three or five years, Weathers would act as Secretary and also use the men employed by the SDC at the gardens in work connected with the flower shows. On 18 January 1904 the Council considered subsidiary questions from SDC, namely, whether the Garden would be open to the public free on Show days and whether the Society expected all the glasshouses to be maintained even if in bad condition. The respective answers were yes, and no. The Society pointed out that, though entry to the gardens would be free, the public would have to pay to visit the Exhibitions. There can be no escaping the irony that no free entry to the Exhibitions was to be given to Stretford residents who would be, in effect, be the owners of the gardens, and that Society subscribers would be free to enter the Exhibitions as usual. There was then no further mention of SDC in the minutes. The explanation for this silence appeared in *The Municipal Journal A Weekly Newspaper,* in January 1904:

A Stretford Poll

A Poll of ratepayers of the district of Stretford was taken on Wednesday on the question of the proposed purchase by the Urban District Council of the Royal Botanical Gardens, Old Trafford. … The result was declared as follows:- For the resolution to purchase the gardens 362, Against 1,350; Majority against, 988.[22]

Above: Longford Park. *Below*: Gorse Hill Park (postcards, n.d.). Stretford already had one park, Victoria Park, opened in 1902 and named in celebration of Queen Victoria's reign. Longford Park was opened in 1911 when the Council purchased the house and garden of John Rylands, wealthy cotton merchant and local Stretford benefactor. In 1923, Gorse Hill Park was established on fields adjoining Talbot Road, close by the site of the Botanic Garden. The postcard shows the entrance gates which formerly graced the entrance to Trafford Park.
AUTHORS'S COLLECTION

The ratepayers were simply not prepared to bear the costs of the enterprise. Other offers to buy were received and came to nothing.

Finally, the Society decided to try to dispose of the Garden to Manchester, and in November 1904 representatives of the Council and the Trustees met with parties from Manchester Corporation with a view to agreeing terms 'as near as possible to those Stretford agreed'. On 12 January 1905 Messrs Bowden and Weathers waited on the Lord Mayor to present the Society's terms. This brief document had only three conditions, the sale of all the land and buildings for £20,000, the Garden to be maintained with respect to its traditional values and the Society to have use of the Garden for exhibitions on 10 or 12 days a year.[23] The Society's accounts showed a loss of £1650 for 1904 so one must assume that they were now determined to sell as quickly as possible. This was confirmed by the meeting of 23 February when the Council decided that 'in the absence of public response to their problems' the Garden would be sold as a whole or, if not, in plots which they would advertise in the local press. Again there was nothing further mentioned in the minutes about the sale of the Garden to Manchester. As with the sale to SDC the information has to be sought elsewhere.

Draft Statement of Terms for sale to Manchester Corporation 12 January 1905 (MBH 2/1/5).

Heaton House,
Lancashire. From S.
Austin, J. Harwood, G
and C Pine, *Lancashire
Illustrated* (London, 1832).

The subscribers were informed of the proposed transfer to Manchester Corporation at the AGM of the Society on 27 January 1905. Mr Horsfall thought it would be 'a fatal mistake if the town did not take possession' but conceded that 'the poorness of attendance seemed to show that the subject was not exciting the community'.[24] Alderman Gibson, too, painted a bleak future, saying 'the circumstances of the Society and of the district had so greatly altered in recent years that it was absolutely impossible for the trustees or committee to keep the gardens'. Somehow there was the aura of regret in these proceedings. The Society knew it must sell the land as a financial necessity yet, in asking to have its own exhibitions, still sought to cling to a last vestige of its glorious, privileged past. There was, however, the Lord Mayor's 'only objection'. The plot of land was not within the city and there was comment that the city had already provided sufficient parks and open spaces for people outside. A letter to the *Manchester Courier* agreed with the Lord Mayor and opposed the purchase, holding up Heaton Park as an example of what to expect. When Manchester had purchased Heaton Hall in 1902 the upkeep was estimated to be £1000 per annum but the previous year, 1904, the costs had in fact been £7000. The minutes of the Manchester City Council meeting on 15 February shows that the purchase of the Botanic Garden was not even referred to the Parks Committee but voted out immediately.[25]

The *Courier* interviewed one of the Society's Trustees the next day and this illuminates the Council's position.[26] The anonymous Trustee claimed that the Society could not be allowed to go further into debt as the liabilities were already £12,000 and the Garden would have to be closed. He added that the Society would leave the gardens closed, probably for 12 years, and then sell the land by which time he estimated it 'would be worth £80,000'. He further explained that the Society had had the restrictive covenants lifted three years previously but needed to wait 15 years from that date before they could sell for commercial purposes. Whatever happened, the money from the sale would not go to the subscribers but would be used for the benefit of horticulture and science. The editorial also expressed strong views on the subject:

> A future generation may go even further in its opinion, and condemn the City Council for its refusal to accept the offer. ... The tide of fashion will surely turn soon in favour of those flower shows which were once the glory of Old Trafford and added another hue to the rainbow of Manchester's not too numerous outdoor pleasures. Gardens so full of memories should not be allowed to fall into the hands of the builder.

In 1905 the proposed advertisement for the sale of the Garden never appeared. A member of the Council, Mr W. T. Rothwell, offered to retire from the Council and take a lease of the gardens for one year. He would maintain them and indemnify the Society against any loss. The offer was accepted unanimously and a tenancy agreement drawn up.[27] No minutes for 1906 appear to have been recorded.

On 5 December 1905, Rothwell agreed to extend his tenancy for further year. When the minutes recommence on 28 February 1907, a draft agreement was being considered between the Society and Messrs Heathcote and Sons (Architects) who wanted to lease the gardens. A voting paper had been sent to all subscribers inviting them to allow a lease for 10 years, at an annual rent of £2000 with the tenants providing a £2000 deposit against dilapidations when the lease was signed. The turn out for voting at the meetings and by post was exceedingly low. There were 15 members at the meeting and, together with the postal votes, voting was 61 votes in favour and one against. The voting paper must also have proposed a resolution permitting the Trustees to obtain a loan of £13,000 to allow the existing mortgage to be paid off using the security of the land, as a second vote was then taken on such a resolution. There were 58 members in favour and one against so both resolutions were carried. The minutes of 27 March 1907 recorded the completion of the legal process and that the lease had been signed with The White City Company Ltd (hereafter WCC): Heathcote had been acting as their agent.[28] The Society had received £2500: the £2000 security and the first quarter's rent. In April the Managing Director of WCC wanted changes; the Whit Week show to take place later than usual, part of the entrance frontage demolished to make new exits and part of the range of glasshouse razed, although in this case, the Council required them to build others of 'equal capacity'.

The Council was still determined to clear the debts though the hoped-for loan was impossible to obtain unless the Trustees would become personally responsible for such borrowings, which they declined to do. New mortgage negotiations began. Weather's position as Secretary was confirmed on 4 July 1907, his terms of employment remained unchanged and he continued to live in the Garden. The WCC wanted to continue to use the Society's medals and the Council again approved; this can only mean that shows were now being organised by the WCC. At the same meeting, the Society's solicitor produced the draft mortgage deed for £13,000, agreed between themselves and the de Trafford Trustees. A special meeting of members was needed to approve the arrangement and voting papers were sent out and agreement was finally given at an adjourned meeting on 20 August with 5 members present and 63 proxy voting papers returned; 61 were in favour so the motion was carried. The mortgage was completed before the next meeting on 5 September so the debts could be cleared.

Trouble was not long coming when the WCC made a late payment of rent in October. Mr J. E. Lees, an Oldham solicitor and a director of the WCC requested the lease be assigned to him and be extended, claiming that his company would pay an extra £200 *per annum* after 10 years. Mr Duckworth disapproved and insisted the work of the Society must be kept separate from the amusement company. The Council thought otherwise and recommended a new lease of 20 years, 10 years at £2000 the rest at £2200 and Mr Lees was to buy all the Society's plants and chattels for £750. Mr Duckworth resigned.[29] To

Royal Botanical and Horticultural Society of Manchester and the Northern Counties.

December 23rd, 1907.

Notice is hereby given, that an EXTRAORDINARY MEETING of Members of the Society will be held in the LORD MAYOR'S PARLOUR, Town Hall, Manchester, on *Monday, December 30th*, at 12 o'clock noon, to approve the following Resolution :—

P. WEATHERS,
Secretary.

RESOLUTION.

"THAT THE AGREEMENT dated the 18th day of December, 1907, and made "between Robert Gibson, Chairman of the Council and for and on behalf of the "Trustees of the Royal Botanical and Horticultural Society of Manchester and "the Northern Counties of the one part, and John Edward Lees of the other part, "being an option for the said John Edward Lees, or a Company to be formed "by him satisfactory as to Capital and otherwise to the Society, to acquire "within two calendar months, a lease of the Botanical Gardens on payment of "the sum of £500 on the signature of the said Agreement, such sum on the "option being duly exercised, to be retained in satisfaction of the first "quarter's rent, and the sum of £750 in purchase of all plants and "chattels on the premises of the Society, and for the right to remove "certain buildings and erections shewn on the plan annexed to the said "Agreement, such lease to be for the term of 20 years determinable by the "tenants at 5, 10, or 15 years at the annual rent payable quarterly in "advance of £2,000 for the first ten years and £2,200 for the remainder "of the term, be and is hereby confirmed And that the Trustees be and "they are hereby authorised and directed on the said option being duly "exercised to execute all such deeds and perform all such acts relating to "the same as may be necessary or proper for carrying the same into effect"

VOTING PAPER.

Members who are unable to attend the Meeting are requested to sign this Voting Paper, marked **X** in either YES or NO column, and return to Mr. P. WEATHERS, Secretary, Royal Botanical Society, Old Trafford.

	YES.	NO.
For the Resolution as above		

*Name*_____

*Date*_____

Notice of Extraordinary Meeting 30 December 1907 and Voting Paper MBH 2/1/5.

authorise the new lease, another notice was sent out to members on 5 December 1907 calling a special meeting for 30 December. The resolution was approved with 45 in favour and 2 against. After 76 years the Society no longer controlled the Manchester Botanic Garden.

'Manchester's "White City": The Botanics Transformed' ran the headline in the *City News* on 2 March 1907, showing a bird's-eye view of the proposed amusement park.

The article claimed that 'prominent showmen' in America and this country were already negotiating for the use of the buildings.[30] The sunken flowerbed, the former fairy fountain, was to become a lagoon and be connected to the original lake by a series of smaller ornamental lakes. Boating was to be an attraction with a depiction of a "Venetian Scene" with gondolas to rent, and

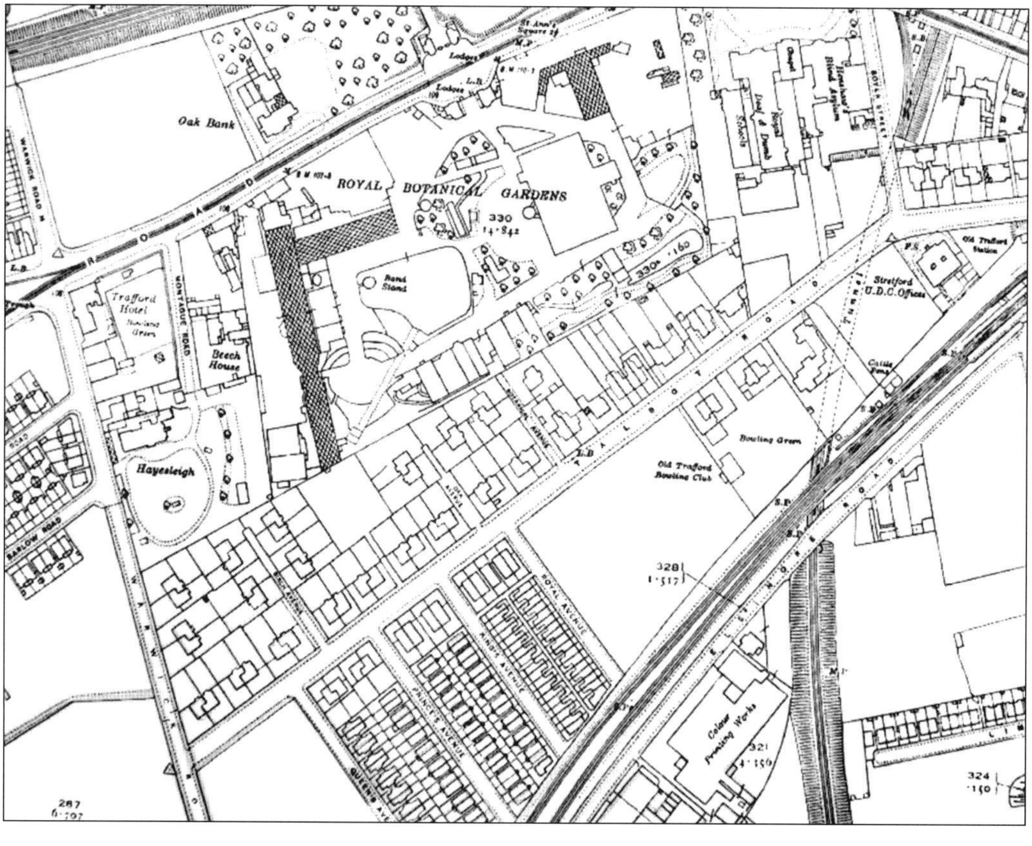

The Botanics Transformed. The White City. *Left*: the water chute; *Right*: inside the ballroom (postcards, n/d.).
AUTHOR'S COLLECTION

The Royal Botanical Gardens in 1913 showing alterations made by The White City C. Ordnance Survey, *Manchester and Salford, Special Edition* (London, 1913).
TRAFFORD LOCAL STUDIES, SALE

An Aside: the fate of the subscription botanic gardens

The movement for subscription botanic gardens flourished in the first half of the nineteenth century, but by the 1850s the Liverpool Garden was owned by the local authority. The other Gardens were being transformed into pleasure grounds with little notice being given to science. By the end of the century Hull's garden no longer existed and the Glasgow and Sheffield gardens also belonged to their respective cities. Liverpool Botanic Garden was sold to Liverpool Corporation in 1847; though it still remains on its site in Waverley it is in need of restoration. Birmingham alone remains on its original site and is run, in conjunction with the City Council, by the original Society.

By the turn of the twentieth century other forms of botanic gardens were appearing. Wealthy industrialists left their gardens to their local universities and these often became the universities' Botanic Garden; examples include Winterbourne, the University of Birmingham Botanic Garden (John MacDonald Nicolson) and Ness Garden, the University of Liverpool Botanic Garden (the Bulley Family). In Manchester, Alderman Fletcher Moss left his private botanic garden in Didsbury to the City who maintains it still as the Fletcher Moss Botanic Garden (2007).

Sheffield Botanic Garden after the restoration funded by The National Lottery.

Liverpool Botanic Garden, Waverly.

the Grand Avenue was to become a Ballroom and a skating rink. The company opened to the public on Whit Monday, 20 May 1907 and the entrance charge was 6d. By evening 25,000 electric lights illuminated the gardens and 32,972 people had passed through the gates,

The Company's *Programme for 1909* claimed the White City was the favoured Amusement Resort of the better class in and around the 'Capital City of the North' and was the best equipped in the country. The changes had been undertaken so that nothing was 'in the least way objectionable to the educated and refined taste' and 'innocent pleasure for the masses and the classes' were being provided. They were exploring other Parks in Europe and America to find further novelties to bring to Manchester.

Notes

1   Minutes, 14 November 1894, MBH 2/1/5.

2   Quoted in Findlay's speech, 'Presentation to Mr. Bruce Findlay', *Gardener's Chronicle,* 19 May 1894, 633.

3   Letter, Bruce Findlay to the editor, 'The Manchester Botanic Society and Allotments', *Journal of Horticulture and Cottage Gardener,* 26 September 1895, 295.

4   See: Allotments, Jellicoe *et al., The Oxford Companion to Gardens*, 9 (Oxford, 1991).

5   Letter, Bruce Findlay to the editor, *ibid.,* 295.

6   Minutes, 11 December 1895, MBH 2/1/5. See also W. T. Tracy and W. B. Pike, *Lancashire at the Opening of the Twentieth Century: contemporary biographies* (Brighton, 1899), for both.

7   Minutes, 8 January 1896, MBH 2/1/5.

8   Minutes, 1 July 1896, MBH 2/1/5 Letters pasted in the minute book are: Letter from Council dated 8 July, 1896 to Florence Findlay, daughter; Letter dated 9 July,1896 in reply.

9   Joseph Broome became a partner in Herman, Samson and Lappoc, Shippers, in 1868. In 1887 he was a founding partner of Broome, Hallworth and Foster, JP for Manchester, 1881 and High Sheriff of Caernarvonshire, 1892–3. Broome was on the committees of several Manchester charities. See W. T. Tracy and W. B. Pike, *Lancashire at the Opening of the Twentieth Century: contemporary biographies* (Brighton, 1899).

10 Minutes, 7 July and 4 August 1897, MBH 2/1/5.

11 Special Meeting, Minutes, 18 January 1898, MBh 2/1/5.

12 Minutes, 6 December 1898, MBH 2/1/5.

13 Minutes, 10 January 1899, MBH 2/1/5.

14 See 'A Greater Hyde Park for a Greater Manchester', *City News,* 8 April 1893.

15 Minutes, 11 March 1899, MBH 2/1/5.

16 'The Manchester Exhibition', *Atlantic Monthly*, November 1857, 33.

17 Minutes, 17 February 1899, MBH 2/1/5.

18 Minutes, 26 June 1900, MBH 2/1/5.

19 It becomes clear in later minutes that the Society was granted a release from its covenants after a 15 year interim period. Minutes, 17 February 1899, MBH 2/1/5. The Deaf and Dumb school bought 4000 square yards from the Society in 1907 for £2856. They bought a further 8000 square yards in 1924.

20 *Ibid.*

21 Minutes, 7 January 1903, MBH 2/1/5. See also Special Meeting and Accounts, 13 January 1903, MBH 2/1/5.

22 *The Municipal Journal A Weekly Newspaper* 13, 22 January 1904.

23 Minutes, 12 January 1905, MBH 2/1/5.

24 'Botanical Gardens Offer The Lord Mayor's View of the "Only Objection"', *Courier,* 28 January 1905. Probably T. C. Horsfall, Life member 1877, a cotton manufacturer and lifelong Manchester philanthropist. He founded the Manchester Art Museum in 1884 for the working classes and the Manchester University Settlement in 1895. His portrait is in the holdings of the Whitworth Art Gallery, Manchester.

25 Manchester City Council Minutes, 15 February 1905 (Manchester, 1905), 234. See also *Appendix to Council Minutes 1904–05, 236–7* (Manchester, 1905).

26 'Royal Botanical Gardens will they be closed? Interview with a Trustee', *Courier,* 16 February 1905.

27 Minutes, 3 and 10 March 1905, MBH 2/1/5.

28   See 'White City Prospectus' (F.4.2.1, Chethams Library, Manchester). The White City Ltd was floated as a company on 16 October 1907. The share capital was £80,000. The Directors were Charles Heathcote, A. J. Bailey, Albert Earwaker, John Edward Lees, James Arthur Tweedale and John Calvin Brown.

29   Minutes, 19 November 1907, MBH 2/1/5.

30   'Manchester's "White City"', *City News,* 2 March 1907. White City Amusement Parks had been established on the East Coast of the United States before coming to Britain.

CHAPTER FOURTEEN

Conclusion

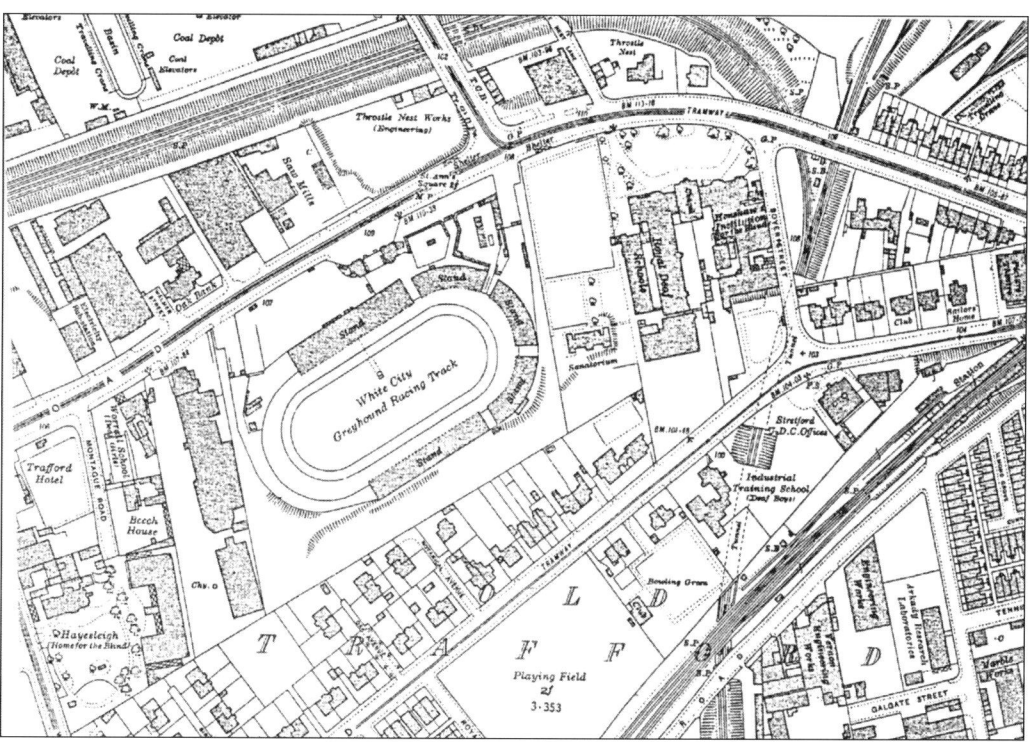

The former site of the Manchester Botanic Garden. Ordnance Survey, *Manchester and Salford* (London, 1931).

After leasing the Garden to the White City Company the Society struggled on. Though the minutes become scrappier and scrappier as the years progressed, they still showed that the finances had not improved. In 1912 the White City Company Ltd went bankrupt and the old order returned. When war was declared in 1914 the Council was still attempting to sell the land. In June 1916 the last exhibition ever to be held in the gardens took place, 'The Garden in Wartime', and lost money again! On the 14 January 1922 the City News published a sad description of the Garden from a correspondent, 'M':

> From the upper deck of a tramcar I was afforded a view of the woeful change that has come over a resort, once so popular with the public. The glasshouses, with their plants and flowers of infinite variety, have gone, as well as other features and all is desolation. ... the next stage will be provided by the interprising [sic] builder. When he appears he will transform the place for business, and the gardens, which were opened in 1831, will become a mere memory.[1]

On 1 November 1927 a special general meeting voted to sell the remaining 11 acres (approx. 4.5 ha; for £40,000 to Canine Sports Ltd for use as a greyhound racing track.[2])

When the Manchester subscription Botanic Garden was founded in 1827 the intention had been to bring to it the latest developments in botany and horticulture and allow members to appreciate science within the precincts of a pleasure ground displaying the beauties of God's benevolence. The financial requirements of maintaining such a Garden in the style expected by the Council and the members were eventually to prove the Society's undoing. Even the upper middle class Council of 1827, which included the Heywood banking family, had little more financial success than the later Councils who were comprised of local tradesmen. Only the drive of Bruce Findlay and his National exhibitions allowed the Society to repay all the debts and own the land at last. By 1890 old problems began to emerge again.

In Manchester, the Botanic Garden initially provided a retreat in the city centre, a *hortus conclusus*, from the smoke, grime and social conditions the Industrial Revolution brought to the city. The exclusivity of the Society was central to this appeal with its blackballing of unsuitable members and the restrictions placed on the lower classes, limiting access to Whit Week and certain exhibitions. Even the attempt made to widen access in 1850 with the opening of public parks, was quickly abandoned in 1853 at the behest of the membership. As this source of income was enjoyed by many of the other subscription botanic gardens, it shows how important was the concept of exclusivity to the Manchester enterprise even at the expense of solving their financial problems.

One undoubted area of success for the Society was in staging shows that drew national acclaim. When Bruce Findlay asserted in 1882 that more than 40,000 or 50,000 people a week visited the exhibitions, he was referring to this area of success. The National Horticultural Exhibitions in Manchester corresponded with the overall growth of such national events. If they can be judged on the pleasure they must have given to their visitors then they were a triumph. Even before the National Horticultural Exhibitions were suggested, the Whit Week opening drew enormous crowds to the Garden from all classes of local society. Certainly, Ann Thompson enjoyed her visit on Whit Monday, 1851:

> My birthday: went by train to the Botanical Gardens – enjoyed them very much, Rain came on before we left. Fought our way home in omnibus.[3]

The rise of the Exhibition as a form of entertainment shows how national trends affected the Garden.

Another national trend that affected the Society was the change in gardening habits in the second half of the nineteenth century. This was the rise of the suburb and the phenomenon known as Villa Gardening when many members of the middle classes moved into the suburbs of Britain's industrial cities to larger houses (villas) and gardens. Gardening and housing in Manchester followed this pattern and the result was that many of the Society's members moved further away from the Botanic Garden. The move to suburbia, where members formed their own horticultural societies and clubs, was one of the reasons the Society failed to attract sufficient members. Perhaps even more important is the philosophical importance of villa gardens, the private middle class *hortus conclusus* in the suburb, a role gardens still play in the technological age of today. This new phenomenon meant the Botanic Garden lost its primary purpose for the members.

But there were other factors. Depressed trade in Manchester and bad weather were often given as reasons for failure by successive Councils and certainly both played an important part. The Garden's location finally also contributed to the failure as growing industrialisation and the opening of Trafford Park added to the Society's problems by making the venue less attractive. Public attitudes were changing; Findlay had complained in court in 1894, 'frivolity has swept the nation'. This was indeed another national trend and by the end of the 1890s many of the outward forms of middle class culture were beginning to decline; Manchester was no exception. In 1880 the Royal Institution had failed and the Corporation stepped in to create the City Art Gallery, a salvation the Manchester Botanic Garden was not to enjoy. The Exchange and the Foreign Libraries were in severe financial difficulties and in 1891 the Portico Library too was lamenting the loss of membership with the treasurer advocating strict economies.

In the event all was not lost with the sale of the Garden. The Society that founded it survived. After the legal progress through the Chancery Court, in 1935 the Royal Botanical and Horticultural Society of Manchester and the Northern Counties became a registered charity and invested the purchase money. The Council, for there are now no subscribing members, continues to administer the income from the investments for the promotion of science and horticulture in Manchester and the northern counties.

On the 150th anniversary of the Royal Botanical and Horticultural Society of Manchester and the Northern Counties in 1977, the Chairman, Mr A. E. R. Goulty, delivered a review of its history to the Tatton Garden Society. He commented that:

> Perhaps one may reach the wry conclusion that the Royal Botanical Society has been more successful in its recent years of gardening by proxy than in the past generations of struggle at Old Trafford.[4]

Does this statement fairly reflect the history of the Society? It is true that, since 1935, the Society has contributed immensely to horticulture in the north of England. The Society continues its work in the horticultural world of Manchester and the northern counties. Beneficiaries have included Manchester University, the Lakeland Horticultural Society (Holehird Gardens), the Northern Horticultural Society (Harlow Carr – now the RHS), the promotion of gardening for the disabled and many local horticultural and botanical societies and students. The Society helped found the Tatton Garden Society

An Aside: The Royal Botanical and Horticultural Society of Manchester and the Northern Counties

The Manchester Botanical and Horticultural Society was founded in 1827. Its purpose was to construct 'a Botanical and Horticultural Garden for Manchester and the neighbourhood'. This was achieved by 1831 when the garden opened in June for the enjoyment of its subscribing members.

The first name change came in 1877 when the Report of the Council for 1876 announced that: 'Her Majesty the Queen has graciously consented to become the Patron of the Society, and on two occasions during the past year has also been an Exhibitor at the Society's Exhibitions held in the Town Hall (Manchester).' The new name was 'The Royal Manchester Botanical and Horticultural Society'.

1881 was the Society's, or rather the Botanic Garden's, Jubilee Year. In August, to celebrate the event, the Society held an International Show. On 27 August the *Gardener's Chronicle* described this magnificent exhibition as 'the greatest combined fruit and flower show ever held in this

The Royal Botanical and Horticultural Society of Manchester and the Northern Counties: Logo of the Society.

country.' In 1883 to honour their own achievement, the Society changed the name again and became 'The Royal Botanical and Horticultural Society of Manchester and the Northern Counties.

The land owned by the Society was finally sold in 1927. The Committee was bound under the terms of the original purchase deed to use any money raised for the good of botany and horticulture. To this end the Society, now consisting of a Committee and no members, became a charity. Under the new rules the main purpose of the Society was, and is, to promote science and art in botany and horticulture.

and is involved with the Quinta Arboretum, a legacy of Sir Bernard Lovell. The Chairman, Mr Goulty, could argue with confidence that post-1935 the Society, untrammeled by the financial burdens of running the Garden, had successfully fulfilled its remit. But surely this should not detract from the work of his predecessors. The endless committees who battled against overwhelming odds, to keep the Garden as both a private garden for their subscribers and a venue where the citizens of Manchester and beyond could come to see the glories of nature and Empire in a beautiful setting, whilst enjoying the exhibitions or promenades.

The Councils of the past, for all their failings, kept faith with the founder's intentions and promoted horticulture and pleasure in the Garden. The Manchester Botanic Society may be a footnote to the history of the region but

Botanical Avenue 2007. One of the last memorials on the site is this cul-de-sac off Talbot Road, leading past an office block, Botanic House, to the site of the back gate into the Garden.

it, like the other subscription botanic gardens, brought the joys of gardening to the attention of an urban population when the industrial city seemed to have obliterated Nature. For that alone the Royal Botanical and Horticultural Society of Manchester and the Northern Counties, and the Garden it founded, deserve public recognition, however belatedly.

Notes

1   'Royal Botanical Gardens', *City News*, 14 January 1922.
2   Minutes, 1 November 1827, MBH 2/1/6. White City went on to be used for dirt track racing, an athletic stadium and the White City shopping centre (current use 2010).
3   Private Diary, Ann Thompson, 7, Heywood Street, Manchester, 9 June 1851 (Thompson family private papers, Manchester). She lived with her brother who was second master at Manchester Grammar School. Ann was well educated, read French and German, played the piano and was interested in literature.
4   'The Botanical and Horticultural Society of Manchester and the Northern Counties', *Tatton Garden Society Quarterly Review, Winter 1977* (Knutsford, 1977), MBH 7/1/33. A. E. R. Goulty was Chairman of both the Royal Botanical Society and the Tatton Garden Society.

Occupational Analysis of the Membership

Members' occupations were found in the Manchester directories of the relevant years using their business or home addresses, where given, in the membership lists of 1833, 1844 and 1877. In the 1877 list, the proportion unclassified is around 25%. This is probably a consequence of members having home addresses in the suburbs and therefore not being listed in the Manchester town directories. (The occupational categories were chosen to correspond with those used for my analysis of the members of the Portico in 1992 so that, although the lists do not correspond in date, some comparisons could be made between the memberships).[1] 'Other' includes a variety of occupations where there were only one or two members, for example architects, builders and teachers. 'Unclassified' in both lists applies to those for whom no occupation was given and none could be found at the business address given. This could mean that they were employees.

Year	1833	1844	1877 A	1877 H	1877 Total
Agent	8 (2%)	16 (4%)	5 (3%)	13 (2%)	18 (2%)
Finance	17 (4%)	15 (3%)	4 (2%)	33 (5%)	37 (4%)
Religion	8 (2%)	7 (2%)	6 (3%)	17 (3%)	23 (3%)
Legal	30 (7%)	30 (6%)	4 (2%)	22 (4%)	26 (3%)
Manufacturer	45 (11%)	28 (6%)	15 (7%)	39 (6%)	54 (7%)
Medical	17 (4%)	12 (3%)	2 (1%)	7 (1%)	9 (1%)
Merchant	74 (17%)	80 (18%)	41 (19%)	113 (19%)	154 (19%)
Textile	127 (30%)	110 (25%)	27 (13%)	125 (21%)	152 (19%)
Women	11 (3%)	25 (6%)	35 (16%)	28 (5%)	63 (8%)
Other	15 (4%)	21 (5%)	20 (9%)	51 (8%)	71 (9%)
Unclassified	70 (16%)	96 (22%)	54 (25%)	155 (26%)	209 (25%)
Total	422	440	214	603	817

Examining the sector for merchants and self-employed members of the Botanical Society as percentages of the overall membership shows changes over the period. In 1833, those engaged in the textile trades formed the largest group; by 1877 merchants had parity with this former majority group. This mercantile group of members, though remaining around 19% over the period of the study, appears to have become an increasingly important sector as the textile membership decreased.[2] This is the group who, it was claimed, knew little about gardening.

Individual professions were low both in numbers and as a percentage of the membership of the Botanical Society but when added together they constituted a large group, 17% in 1833, 14% in 1844 and 11% in 1877.[3] Members of the legal

Occupation	1844	1877 Annual	1877 Hereditary	1877 Total
Architect	2 (0.45%)	2 (1%)	5 (0.8%)	7 (0.85%)
Brewer/Wine Merchant	11 (2.5%)	5 (2%)	10 (1.6%)	15 (1.8%)
Victualler / Hotelier	5 (1.1%)	0 (0%)	11 (1.8%)	11 (1.4%)
Engineer	3 (0.6%)	3 (0.5%)	17 (2%)	20 (2.5%)
Estate Agent/ Valuer	4 (0.9%)	3 (0.5%)	12 (2%)	15 (1.8%)
Merchant (Unspecified)	17 (3.9%)	14 (2.3%)	12 (2%)	26 (3.2%)
Merchant – Building Trades	12 (2.7%)	3 (0.5%)	9 (1.5%)	12 (1.5%)
Merchant – Clothing	4 (0.9%)	5 (2%)	6 (1%)	11 (1.4%)
Merchant – Coal, Oil, Tallow	3 (0.6%)	4 (1.9%)	8 (1.3%)	12 (1.5%)
Merchant – Provisions	15 (3.4%)	10 (4.7%)	9 (1.5%)	19 (2.3%)
Merchant – Furnishings	4 (0.9%)	6 (2.8%)	9 (1.5%)	15 (1.8%)
Merchants – Pharmacists	0 (0%)	0 (0%)	2 (0.3%)	2 (0.24%)
Nurserymen/ Landscaper	2 (0.45%)	1 (0.5%)	6 (1%)	7 (0.9%)
Self-employed builder, joiner	4 (0.9%)	0 (0%)	3 (0.5%)	3 (0.37%)
Other	0 (0%)	3 (0.5%)	5 (0.83%)	8 (1%)

profession constituted 6% in 1833 but declined to 3% in 1877 although over this period lawyers were nationally developing their status as professionals.[4] Medical members were always a small percentage of the Botanic Society membership though, given that botanic gardens and medicine had strong historic links, more might have been expected: 4% in 1833, declining to 1% in 1877.[5] It could be that lawyers and doctors were groups who at the beginning of the nineteenth century wanted to improve their social and professional standing and sought membership of organisations perceived to carry this advantage.[6] The percentage of members connected to the financial sector also remained low at all dates examined, averaging 4%.[7] Perhaps an organisation, where meeting those who could advance their careers might be chance encounters, was not as attractive as other organisations in Manchester, for example the Portico Library.

An analysis of the membership of the Botanical Society Council for the selected years, 1845–6, 1853–4 and 1877 shows that they do not reflect the overall membership in terms of their occupational social groups. The Council for 1845–6 shows professionals constituted 27% though they formed only 17% of the membership. An almost identical pattern was seen for 1853–4 even though professionals had fallen to 14% of the membership. By 1877 when their overall numbers had fallen to 11%, professionals made up 22% of the Council, thus making them the second largest grouping. In fact in 1845 and 1853 the professionals on the Council formed a larger group than the merchants and those connected to the textile industry. By 1877 members of the Council connected to the cotton trade were in the majority (33%), a position they also held in the total membership (19%). These results seem to confirm the earlier speculation that belonging to Manchester Societies, here the Botanical Society, was seen as a way to advance the emerging professions of law, medicine and finance. The further conclusion is that such advancement was better realised by becoming actively involved in the management of the Society than by merely becoming a member.

The Botanical Society appealed specifically to women in 1827 by providing a garden in which to appreciate botany and a venue for pleasurable walks. As women's membership of organisations in their own right was unusual they were designated a category in the occupational analysis.

Date of list	1833	1844	1877	1877 Annual	1877 Hereditary
Women	11 (3%)	25 (6%)	63 (8%)	35 (16%)	28 (5%)

The most interesting fact appears in the 1877 list. The women were split between the annual and hereditary members: 35 annual and 28 hereditary, together 8% of the total membership for the year.[8] Women made up 16% of the total annual members but only 5% of the total hereditary members. Four women annual members were single and, of the 29 married annual members, two had male relatives living at the same address who were hereditary members.[9] In 1877 only one woman annual member listed an occupation: Mrs Sherwood, Old Trafford was a glass and china merchant (no woman gave an occupation in earlier lists). Of the women hereditary members seven were single and four of these had the same listed address as a male hereditary member. Two women hereditary members gave occupations: Mrs Pike of Old Trafford listed her occupation as a manufacturer of cap and hat trimmings and Mrs Cole gave her address as The Nursery, Didsbury, a well-known local plant nursery.

Though this is a very small survey some occupations do show significant changes. Engineers, for example, show almost 2% growth in membership between 1844 and 1877 and this may reflect the significant growth in Manchester of the engineering sector. The national trend in the growth of cities as retail centres with the establishment of large department stores selling household goods and ready-to-wear clothing is reflected in the growth in members with occupations connected to the retail sector.[10] This demand can also be related to the buying power of an expanded middle class and the growth of suburban housing. The increase in nurserymen and landscapers as a percentage of the membership is less surprising, as the increase in personal gardens led to greater demand for services connected to horticulture, and the Society offered an ideal place to meet potential clients.[11]

There were two important characteristics shared by the memberships. All male members had middle class status and most were engaged, in their various ways, with Manchester's trading businesses and the creation of the city's wealth; the growth of certain groups of members, for example engineers, professional gardeners and members of the financial industry, reflected the growth of these sectors in Manchester and district.

The analysis of the addresses of the women members in 1877 showed that the majority of annual members lived near the garden. Only one hereditary member did so. The remaining hereditary women members lived in the suburbs, confirming that the wealthy hereditary families were the ones who had fled the city centre, away from the dirt and industry.

Notes

1   A. Brooks, *'A Veritable Eden', The Manchester Botanic Garden and the Movement for Subscription Botanic Gardens*, PhD Thesis, University of Manchester, 2007; also A. Brooks, *The Portico Library and Newsroom*, (MA, Manchester Polytechnic, 1992). The Portico lists used, from the Portico Archives, were for 1806, 1853 and 1903. The analysis showed that, as may be expected, the common link between the two organisations was the social status of the members. Unlike the Botanical Society, membership of the Portico was never by ballot of potential members and its exclusivity was maintained by the cost of the shares; in 1806 the subscription was a guinea a year with an initial fee of 13 guineas. Membership was ecumenical and discussion of politics and religion was prohibited. See also A. Brooks and B. Haworth, *Portico Library A History*, 6–7 (Manchester, 2000)

2   S. Gunn, *The Public Culture of the Victorian Middle Class* (Manchester, 2007); S. Gunn and R. Bell, *Middle Classes: their rise and sprawl* (London, 2002)

3   For this discussion the professions comprise medicine, religion, law and finance

4   V. R. Parrott, *Pettyfogging to Respectability. A History of the Development of the Profession of Solicitor in the Manchester Area 1800–1914* (PhD Thesis, University of Salford, 1992)

5   W. J. Elwood and A. F. Tuxford, *Some Manchester Doctors A biographical collection to mark the anniversary of the Manchester Medical Society 1834–1984* (Manchester, 1984)

6   W. J. Reader, *Professional Men: the rise of the professional classes in nineteenth-century England* (London, 1966)

7   L. H. Grindon, *Manchester Banks and Banking* (Manchester, 1877); also Kidd, *Manchester*, 107–8 (Keele, 1993); A. Brooks and B. Haworth, *Boomtown*, 48–53

8   Whether this was common to all years after annual membership began cannot be established

9   There is significance to this. Family membership included children. The age limit for sons was 18 but there was no age limit for daughters. The family relationship for these female members is therefore open to speculation

10 A. Adburgham, *Shops and Shopping 1800–1914* (London, 1964); Kidd, *Manchester*, 190–1 (Keele, 1993)

11 H. Pryor and J. Hitchens, *Altrincham Gardener's Societ; a hundred years of professional gardeners* (Timperley, 2003)

The Herbarium of the Manchester Botanical and Horticultural Society

CXCIX. THE MOSS FAMILY.—*Mus'ci.*

The Mosses, in their sweet delicacy and elaborate organisation, prefigure all the most pleasing phenomena that pertain to flowering-plants. Every idea that we find culminating either in the wild forest or in the

Fig. 215.
Leskea polyantha.

garden, is anticipated ; yet on a scale so minute that without a microscope it is impossible to discover it. The foliage and general figure of the larger species may be observed with ease ; but the fructification, on which the distinctions of genera and species primarily depend, and in which the rare and inexpressible beauty of these little plants is chiefly manifest, is in no case to be made out accurately until magnified, as

Leskia polyantha. From L. H. Grindon, *British and Garden Botany* (London, 1864).
AUTHOR'S COLLECTION

The Manchester University Herbarium is part of the Manchester Museum and one of the country's most important collections resulting from the amalgamation of several herbaria, both private and corporate.[1] A herbarium became part of the University (then Owen's College) when the collections of the Manchester Natural History Society were given to the infant college in 1868 when the Society closed.[2] In 2005, the Herbarium of the Manchester Botanical and Horticultural Society was discovered at the University Herbarium after opening several previously uncatalogued boxes for classification.[3] The source given for the specimens is of special interest and though incomplete shows that some came from Kew, Liverpool, Hull and Manchester Botanic Gardens (www. museum.manchester.ac.uk/ourcollections).

Like the Natural History Society, the Manchester Botanical and Horticultural Society also had a Herbarium. It is clear from the surviving minutes of the Botanical Society that a Library was established in 1830 and housed in a room in the gatehouse above the Council room.[4] No mention of the Society's Herbarium was made in the minutes until 1845 when the Council approved plans for a new house to be built to accommodate it.[5] The new building was required as there was insufficient space in the current room to display the new *Hortus scions* donated by a Dr Fleming, together with the Herbarium of the late Mr Hobson already in the possession of the Society. It would seem probable therefore that the Herbarium had been displayed in the Library until 1845. The only other mention of the Herbarium was in the *Report of the Council* for 1854 when it was reported that the Herbarium was to be open for study not only to members but also to young men employed in the Garden.[6]

A starting point for seeking further information was to find material on the Herbarium of Mr Hobson. Edward Hobson, born in Ancoats in 1783, was one of Manchester's artisan naturalists, and a 'Memoir' of Hobson was read to the members of the Manchester Literary and Philosophical Society on 19 February 1839.[7] This biography reveals that Hobson's special interest was mosses and by 1816 he was exchanging moss specimens with William Hooker, then in Glasgow.[8] Hobson became first President of the Banksian Society of Manchester in 1828 and the following year assisted in the establishment of the museum of the Manchester Society for the Promotion of Natural History (later the Natural History Society). Hobson retired to Bowdon, Cheshire, for his health in early 1830 and died there in 7 September 1830. A letter from Hooker to John Hampson of Manchester (a member of the Manchester Botanical and Horticultural Society) on 2 October 1830 describes Hobson's book on mosses, *A Collection of Specimens of British Mosses and Hepaticae etc.*, as containing 'beautifully preserved specimens themselves', in effect a herbarium as a book.[9] Hooker then assures Hampson that if the Botanical Society were to raise funds to purchase Hobson's collection he would subscribe £5. Moore stated that Hobson's Herbarium was acquired by the Botanical Society and deposited in the Library at the Garden (Hobson's insect collection went to the museum at the Mechanic's Institute) and it was Hobson's Herbarium that was referred to in the minutes of 1845.[10]

When Campbell was dismissed as Curator in 1857 a new post was created for him as Curator of the Museum. This was formed in 1858 when the minutes of 7 May record that Council member Mr Hibberd was to visit Sir William Hooker at Kew to ask for his co-operation in the establishment of the Museum as it was to be modelled on the botanical museum there.[11] The Museum appears to have been a continuation of the Herbarium and when Campbell was dismissed in 1862 the Museum, too, disappears from the minutes.[12] It seems unlikely that such a valuable collection would simply have been destroyed and therefore the assumption must be that the Society donated the collection to another, probably local, organisation. The obvious candidate would have been the Manchester Natural History Society as it already had a Herbarium, which had been created in 1860 by the amalgamation of several of its collections.[13] There appear to be no records extant that could confirm this supposition.

Though the 2005 discovery confirmed that the Herbarium of the Manchester Botanical and Horticultural Society was now part of the collection at the Manchester Museum, a link between the Botanical Society and the Manchester Natural History Society remained tenuous. However the Manchester Museum Herbarium does have two copies of Hobson's *A Collection of Specimens of British*

Mosses and Hepaticae etc. One of these, the complete three volumes, bears the inscription 'presented to the Manchester Museum by Alderman King'. The second, which is volume one only, has the stamp of the Manchester Natural History Society inside the cover. There was no suggestion in Moore's article that a copy of this rare book was to be found at the Natural History Society. I contend therefore that this is part of the copy from the Library at the Botanical Society, stamped on its accession to the Natural History Society in 1862 (with the Herbarium from the Botanic Garden) and passed to Owen's College with the rest of the Natural History Society's Herbarium in 1867.

CC. THE MARCHANTIA FAMILY.—*Hepat'icæ.*

Minute plants, with usually much the aspect of mosses, but distinguished from them by their thecæ opening into four or about eight valves; by the absence of a calyptra, and usually of a columella; and by the frequent presence of spiral chains among the spores. Like the mosses, they have their habitats in moist and shady ground, near

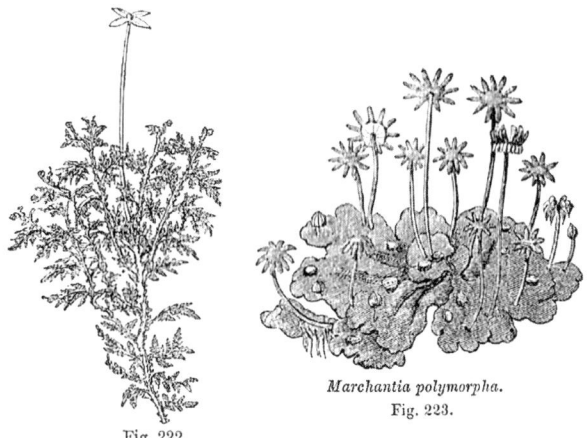

Fig. 222.
Jungermannia tomentella.

Marchantia polymorpha.
Fig. 223.

Mosses *Jungermannia tomentella* and *Marchantia polymorpha*. From L. H. Grindon, *British and Garden Botany* (London, 1864).

rivulets, on moist banks, among moss in woods and cloughs, and upon trees; often also upon rocks, where they are exposed to incessant trickling and dropping of water, or to the spray of little cascades. In colour they are either green or purplish. They have a distinct axis of growth, and are usually provided with distinct and symmetrical, though very minute, leaves, which being imbricated in opposite ranks, give a flattened character to the branches;—sometimes, however, stem and

Notes

1   Herbarium History. *The Manchester Museum: Botany*, 17 (Manchester, 2004). See also J. W. Franks, *HERB MANCH. A guide to the contents of the Herbarium of Manchester Museum*, Manchester Museum Publications New Series 1973; www.museum. manchester.ac.uk/ourcollections (2007).

2   W. E. A. Axon, *Annals of Manchester*, 168 (Manchester, 1886). See also D. E. Allen, *The Naturalist in Britain: A social history* (London, 1976) and *The Botanists: A history of the Botanical Society of the British Isles through a hundred and fifty years* (Winchester, 1986).

3   The list of contents can be found at www.herbariaunited.org (2007).

4   Minutes, 9 June and 30 June 1830, MBH 2/1/1.

5   Minutes, 10 September 1845, MBH 2/1/2 There were no entries to confirm this house was built.

6   *Report of the Council of the Manchester Botanical and Horticultural Society to the Twenty-sixth general annual meeting of the Members, held in the Town Hall, Manchester, on Monday, 6th of March, 1854* (Manchester, 1854), MBH 3/2/17

7   J. Moore, 'A Memoir of Mr. Edward Hobson, author of Musci Britannici, &c.', *Memoirs of the Manchester Literary and Philosophical Society* 11, Second Series 6, 1842, 297–324. See also J. Cash, *Where there's a will, there's a way! or Science in the cottage. An account of Naturalists in Humble Life*, 41 (London, 1873).

8   J. Moore, 'A Memoir of Mr. Edward Hobson, author of Musci Britannici, &c.', *Memoirs of the Manchester Literary and Philosophical Society* 11, Second Series 6, 1842, 309. Also amongst Hobson's correspondents was George Cayley, a fellow Mancunian, who had travelled to Australia in 1800 as a plant collector for Joseph Banks. In 1815 Caley was appointed Superintendent of St Vincent Botanic Garden from where he sent Hobson specimens of tropical plants and ferns.

9   *Ibid.*, 322. The title in full is *A Collection of Specimens of British Mosses systematically arranged with Reference to the Musicologia Britannica, English Botany, and British Jungermannia,* in 3 Vols (Manchester, 1822). The book was intended as a companion to William Hooker and Thomas Taylor's *Muscologia Brinnica: containing the Mosses of Great Britain and and Ireland* (London, 1818), in which Hobson's book was advertised.

10   Minutes, 10 September 1845, *ibid.*

11   Minutes, 7 May 1858, MBH 2/1/4. Lists of donors appear in the minutes and included vegetable substances duplicates from Kew, seeds from the Calcutta Botanic Garden and sugar cane (*Saccharum Violacum)* from Cuba. New plants were obtained as offsets from the canes donated by member Joseph Hanson on his return from the West Indies and some were sent to the Museum. A letter was sent to James A. Turner, MP (a member) asking if he would request specimens from the East India Company. Donations were also received from The Cotton Supply Association and Messrs. Fry, Union Street, Bristol.

12   Minutes 16 April 1862. Campbell was dismissed with effect from 29 September 1962 and released from duties on 1 July 1862.

13   'Herbarium History. *The Manchester Museum: Botany*, 17 (Manchester, 2004).